Manual of Methods for
Marine Plankton

The Authors

Dr. R. Santhanam is the former Dean of Fisheries College and Research Institute, Tamilnadu Veterinary and Animal Sciences Univeristy, Thoothukudi, India. His fields of specialization include Fisheries Environment and Marine Biology. He is presently serving as a Consultant for Advanced Aquatic Environmental Research Services (AAERS), Sultanate of Oman and Fisheries Expert for various Government and Non-governmental Organisations of India. Dr. Santhanam has published 10 books on various aspects of Fisheries Science and 70 research papers. He was a Member of American Fisheries Society, World Aquaculture Society and Global Fisheries Ecosystem Management Network, United States and IUCN's Commission on Ecosystem Management, Switzerland.

Dr. A. Srinivasan is presently working as Professor and Head, Department of Fisheries Environment, Fisheries College and Research Institute, Tamil Nadu Fisheries University, Thoothukudi, India. His fields of specialization include Marine Plankton Dynamics and Marine Pollution. He has guided 6 postgraduate students for their thesis research work and served as Advisory Committee Chairman/Member for 15 postgraduate students. To his credit, he has 40 research papers published in peer- reviewed national and international journals and a book. On invitation, he has attended the IV International Crustacean Congress held in Amsterdam, The Netherlands during 1997. Under India-FAO, AHRDP programme, he has undergone a 3 month training programme on Fisheries Management at Texas A&M University, USA. He has also served as an Academic Council Member of Tamilnadu Veterinary and Animal Sciences University, Chennai and he was the Principal /Co-Principal Investigator for 7 externally funded research projects.

Dr. S. Ajmal Khan is presently serving as Professor Emeritus, Centre of Advanced Study in Marine Biology, Annamalai University, India. His fields of specialization include Marine Fisheries, Biodiversity Assessment and Crustacean Larval Ecology. He was invited by King Fahd University of Petroleum and Minerals, Kingdom of Saudi Arabia to participate in a project sponsored by the U.N. on Damage analysis in the Arabian Gulf (Persian Gulf). He has served as a member in the Thematic Group on Biodiversity of the Union Ministry of Environment and Forests, Academic Council of Tamil Nadu Veterinary and Animal Sciences University and Research Advisory Council of Central Marine Fisheries Research Institute, Indian Council of Agricultural Research. He has published several research papers in nationally and internationally reputed journals.

Manual of Methods for
Marine Plankton

R. Santhanam

A. Srinivasan

S. Ajmal Khan

2015

Daya Publishing House®

A Division of

Astral International Pvt. Ltd.

New Delhi – 110 002

Cataloging in Publication Data—DK
Courtesy: D.K. Agencies (P) Ltd. <docinfo@dkagencies.com>

Santhanam, Ramasamy, 1946-
Manual of methods for marine plankton / R. Santhanam, A. Srinivasan, S. Ajmal Khan.
p. cm.
Includes bibliographical references (p.   ) and index.
ISBN 9789351305576 (International)

1. Marine plankton. 2. Marine phytoplankton. 3. Marine zooplankton. I. Srinivasan, A., 1963-, joint author. II. Ajmal Khan, S., joint author. III. Title. IV. Title: Marine plankton.

DDC 578.776   23

Published by	:	**Daya Publishing House®**

Published by : **Daya Publishing House®**
A Division of
Astral International Pvt. Ltd.
– ISO 9001:2008 Certified Company –
4760-61/23, Ansari Road, Darya Ganj
New Delhi-110 002
Ph. 011-43549197, 23278134
E-mail: info@astralint.com
Website: www.astralint.com

Laser Typesetting : **Classic Computer Services**, Delhi - 110 035

Printed at : **Replika Press Pvt. Ltd.**

PRINTED IN INDIA

Preface

Marine plankton, the diverse group of drifiting organisms are classified into phytoplankton and zooplankton. They assume greater significance as the fishery potentials of seas and oceans largely rely on their density and distribution. This book titled Manual of Methods for Marine Plankton has been prepared to supplement the requirements of the Marine Science students and researchers in the methodologies of marine plankton. This publication covers several topics such as methods of collection, narcotisation, fixation and preservation, estimation of biomass, staining and mounting techniques, micrometry and size measurements and species diversity besides a detailed account of different groups of phytoplankton and zooplankton (holoplankton and meroplankton).

We strongly hope that the use of this book would certainly enhance the reader's skill in marine plankton research in relation to sustainable fisheries.

The authors convey their sincere thanks to Dr. N. Jayabalan, Fisheries Advisor, Advanced Aquatic Environmental Research Services (AAERS), Sultanate of Oman for his suggestions on the manuscript. The secretarial assistance offered by Mrs. Albin Panimalar Ramesh is gratefully acknowledged.

R. Santhanam
A. Srinivasan
S. Ajmal Khan

Contents

Preface   *v*

1. **Introduction**   **1**

    1.1. Plankton by Size   1
    1.2. Importance of Phytoplankton   3
    1.3. Importance of Zooplankton   3
    1.4. Importance of Plankton Study   4
    1.5. Threats to Plankton   4

2. **Methods of Collection**   **5**

    2.1. Collection of Nanoplankton   5
    2.2. Collection of Microplankton   9

3. **Plankton Processing Procedures**   **17**

    3.1. Fixation and Preservation of Phytoplankton   17
    3.2. Zooplankton Specimen Processing Procedures   18
    3.3. Bottling and Labelling   20

4. **Staining and Mounting Techniques**   **21**

    4.1. Mounting of Phytoplankton   21
    4.2. Staining, Dissection and Mounting of Zooplankton   22

5. **Estimation of Biomass (Standing Stock)**   **23**

    5.1. Volumetric Methods   23
    5.2. Gravimetric Methods   24
    5.3. Chemical Method   25
    5.4. Quantitative Analysis   25

6. Estimation of Chlorophylls   **31**

6.1. Determination of Chlorophyll *a, b* and *c* in Spectrophotometer   32

6.2. Estimation of Chlorophyll *a* and Phaeophytin *a*   33

6.3. Estimation of Plant Carotenoids   33

7. Production of Plankton   **34**

7.1. Estimation of Primary Production   34

7.2. Estimation of Secondary Production   37

8. Micrometry and Size Measurements   **39**

8.1. Ocular and Stage Micrometers   39

8.2. Calibration of Ocular Micrometer   41

8.3. Size Determination   42

9. Species Diversity   **43**

9.1. Species Diversity Indices   43

9.2. Similarity Index   44

9.3. Coefficient of Community   45

10. Identification of Phytoplankton   **46**

10.1. Diatoms   46

10.2. Dinoflagellates   72

10.3. Other Groups   84

10.4. Toxic and Harmful Species   88

11. Identification of Zooplankton–Holoplankton   **102**

11.1. Protozoans   102

11.2. Cnidarians (Coelenterates)   122

11.3. Acnidarians   130

11.4. Marine Rotifers   131

11.5. Planktonic Annelids   133

11.6. Arrow Worms (Chaetognatha)   136

11.7. Cladocerans   136

11.8. Ostracods   139

11.9. Amphipods   142

11.10. Copepods   143

11.11. Mysids   167

11.12. Decapods   167

11.13. Molluscs   168

11.14. Protochordates   173

12. Identification of Zooplankton–Meroplankton   **178**

    12.1. Larvae of Cnidarians   178

    12.2. Larvae of Planktonic Flatworms   179

    12.3. Larvae of Nemertine Worms   180

    12.4. Larvae of Brachiopods   181

    12.5. Larvae of Phoronids   182

    12.6. Larvae of Bryozoans   182

    12.7. Larvae of Crustaceans   183

    12.8. Larvae of Molluscs   195

    12.9. Larvae of Echinoderms   199

    12.10. Larvae of Protochordates   204

    12.11. Eggs and Larvae of Fishes (Ichthyoplankton)   206

Literature Cited   *211*

Index   *213*

Chapter 1
Introduction

The term 'Plankton' - (Greek-wandering or drifiting) was coined by Victor Hensen in 1887. The plankton are found all over the planet, *i.e.* from the polar-regions to the tropics, freshwater lakes to the sea. They are mostly small organisms, drifting along with the currents and unable to swim against the waves. These small creatures at the bottom of the food chain give life to most of what one sees in the sea. Based on nutrition, plankton are classified into phytoplankton (autotrophs) and zooplankton (heterotrophs). Based on the duration of planktonic life, they are either Holoplanktonic *i.e.* planktonic throughout the life (*e.g.* copepods, cladocerans, chaetognaths, pteropods, etc.) or Meroplanktonic *i.e.* plankonic for a portion of life (temporary plankton) (*e.g.* Larvae of benthic invertebrates and fish larvae (ichthyoplankton).

1.1 Plankton by Size

☆ Megaplankton, Over 20 mm

☆ Mesoplankton, 0.2 mm - 2 mm

☆ Microplankton, 20 μm - 200 μm

☆ Nanoplankton, 2 μm - 20 μm

Functional Groups of Marine Phytoplankton

Class	Groups
Bacillariophyceae	Centric and pennate diatoms
Dinophyceae	Dinoflagellates
Cyanophyceae	Blue-green algae (Cyanobacteria)
Cryptophyceae	Brownish-green algae
Raphidophyceae	Chloromonads
Dictyochophyceae	Silicoflagellates

Contd...

Contd...

Class	Groups
Prymnesiophyceae (Haptophyceae)	Coccolithophorids
Euglenophyceae	Flagellates
Prasinophyceae	Flagellate green algae
Chlorophyceae	Green algae

Functional Groups of Marine Zooplankton

Kingdom	Phylum	Zooplankton Taxa (Class or Order)	Description
Protozoa	Protozoa	Foraminifera	Single-celled, shelled amoeba
	Actinopoda	Acantharia	Radiolarian protozoa
	Retaria	Polycystinia (rads)	Radiolarian protozoa
	Cercozoa	Phaeodaria (rads)	Shelled amoeba protozoa
	Ciliophora	Ciliata (aloricate)	Aloricate ciliates
		Tintinnina	Tintinnids (loricate ciliates)
Animalia	Cnidaria	Hydrozoa	Hydroid, jellyfish
		Siphonophora	Colonial jellies
		Cubomedusae/ Cubozoa	Box jellyfish
		Scyphomedusae/ Scyphozoa	True jellyfish
	Ctenophora	Ctenophora	Comb jelly
	Rotifera	Rotifera	Tiny, multicellular, wheel of cilia around mouth
	Platyhelminthes	Platyhelminthes	Flat worms
	Nematomorpha	Nectonema	Horsehair worms
	Nemertea	Nemertinea	Ribbon worms
	Annelida	Polychaeta	Segmented worms
	Mollusca	Heteropoda	Laterally compressed pelagic snails
		Pteropoda	Sea butterflies
		Nudibranchia	Sea slugs
		Cephalopoda	Squid, octopods
	Arthropoda-	Cladocera	Water fleas
	Crustacea	Ostracoda	Seed shrimp
		Isopoda	Isopods, pill bugs
		Copepoda	Copepods
		Mysidacea	Mysids
		Amphipoda	Shrimp-like crustacean
		Euphausiacea	Krill
		Decapoda	Shrimp

Contd...

Contd...

Kingdom	Phylum	Zooplankton Taxa (Class or Order)	Description
	Chaetognatha	Chaetognatha	Arrow worms
	Chordata	Appendicularia	Larvaceans
		Pyrosoma	Free-floating, ciliated, biolumine-scent tunicates
		Doliolida	Barrel-shaped tunicate
		Salpida	Barrel-shaped tunicate

1.2 Importance of Phytoplankton

Phytoplanktonic photosynthesis accounts for roughly half of the primary productivity on earth and plays an important role in the ocean's carbon cycle. As photosynthesis occurs in phytoplankton, carbon dioxide is incorporated into the cells and taken out of the environment. During this process, more than 100 million tons of inorganic carbon is fixed each day around the world, reducing the amount of carbon dioxide in the atmosphere.

This massive conversion of inorganic carbon into a useable form makes much of the life in the oceans to survive. Carbon gets converted into sugars, which are stored in the cells. These sugars are then eaten by zooplankton, filter feeders, and baleen whales. Zooplankton are eaten by small fish species, and which are further consumed by salmon, tuna, seabirds, marine mammals and all throughout the food web. Phytoplankton is in such high demand that the entire phytoplankton biomass of the world's oceans is consumed by filter feeders, from barnacles to baleen whales, every 2 to 6 days.

1.3 Importance of Zooplankton

The zooplankton are the second step in the marine food chain. That is, they eat the phytoplankton and are themselves eaten by small fishes. A total of approximately 7,000 species in 15 phyla have been described so far. Certain mesoplankters, particularly copepods and cladocerans are essential as food for early fish larvae and for larger predacious zooplankters, which in turn are fed upon by late larval and postlarval fish and other organisms. In estuaries, macroplankters such as mysid shrimp and gammarid amphipods may be the most important food chain link in habitats bounded by extensive salt and brackish marshes, which themselves often are important fish nursery grounds. Another important aspect of zooplankton behaviour is the periodic vertical migration exhibited by many copepods. The most common pattern is to migrate deeper in the water column during daytime and ascend towards the surface at night. The diel (or daily) vertical migration of many planktonic organisms may be influenced by the abundance of both food items and predators, as well as other environmental factors such as light, salinity, and temperature.

Marine zooplankton are important indicators of environmental change associated with global warming and acidification of the oceans. They are also significant mediators of fluxes of carbon, nitrogen, and other critical elements in ocean biogeochemical cycles. A global-scale baseline assessment of marine zooplankton biodiversity is critically needed to provide a contemporary benchmark against which future changes in seas and oceans can be measured.

1.4 Importance of Plankton Study

By examining patterns in plankton distribution one can learn what effect climate change is having on marine ecosystems. Since plankton are not harvested or exploited like fish or intertidal organisms, adjustments in distribution and abundance may be attributed to changing environmental factors. As plankton are indicators of healthy aquatic environments, long-term studies have been carried out on plankton since the 1930s with numerous research projects continuing today.

Warming of the world's oceans has already caused major shifts in plankton distribution and abundance on a global scale. From 1999 to 2004, when a Sea Surface Temperature (SST) change exceeding ± 0.15°C occurred in 74 per cent of the globe's oceans, a decrease in plankton productivity in these areas was also observed, effectively decreasing the amount of available energy in the food chain.

It has been reported that the earth's oceans have been warming for the past 40 years, and warming is going well below the surface to at least 3000 meters. Warming waters may exert impact on many phytoplankton species that live in the top layers in order to carry out photosynthesis, and studies have shown that when phytoplankton populations suffer, so do other species in the surrounding areas.

1.5 Threats to Plankton

Warming water is not the only threat to plankton. As atmospheric carbon dioxide (CO_2) level increases, the oceans absorb more of this gas. It has been estimated that since the year 1800, the oceans have taken up roughly 120 billion metric tons of human generated CO_2. Currently, the oceans are up-taking roughly 20-25 million tons each day. As the oceans take in carbon dioxide, the gas forms carbonic acid, lowering the pH of the ocean water and turning it dangerously acidic. As this acidification occurs at the rate 100 times faster than ever recorded, it is estimated that by the end of the 21st century, the surface waters in some of the world's oceans may not be able to support plankton in general and shell-bearing plankton in particular.

Historical evidence has shown that plankton do not recover easily from catastrophes. When a population crash occurred across the oceans 65 million years ago, it took approximately 3 million years for the plankton to recover. How the plankton around the world will be affected in the long-term by abrupt climate change is difficult to predict. It is therefore very clear that with less plankton, ecosystems from the poles to the tropics, and from freshwater to the salty seas will be negatively impacted.

Chapter 2
Methods of Collection

The sampling strategy of plankton adopted depends on the purpose of the study. Sampling may be for qualitative analysis, in which case, the main objective is to obtain a rapid picture of species present or quantitative, in which case, the objective is to determine the abundance of different species of plankton.

The methods of collecting different fractions of plankton such as nano and microplankton involve collecting the water in bottles, water samplers or by pumps and filtration of water by suitable net cloth. The plankton collections are normally made by horizontal, oblique and vertical hauls. In the horizontal sampling, the net is towed at a slow speed usually for 5 to10 minutes. The towing speed of the net should be such that the maximum amount of water enters through the mouth of the net for better filtration and gear used can withstand the strain. The towing speed of the net recommended for horizontal sampling is 1.5 to 2.0 knots. Most of the plankters are known to migrate vertically in response to light conditions. Hence, their occurrence is normally poor in the upper layers during daytime. For better quantitative and qualitative plankton collections, the suitable time for horizontal plankton sampling would be before dawn, after dusk or night. The horizontal collections are mostly carried out from the surface and subsurface layers. The vertical haul, on the other hand, is made to sample the water column. During this sampling, the net is lowered to the desired depth and hauled slowly upwards. The plankton sample collected is from the water column traversed by the net. Closing mechanisms are used to sample a specific body of water. The samples taken with closing nets are analyzed to study plankton abundance at different depths.

2.1. Collection of Nanoplankton

Bottle/Water Samplers

Sampling by water sampler helps to obtain a correct picture of the quantitative composition of all the fractions of plankton. These are ideal devices for quantitative plankton collections, as required quantities of water can be collected from the desired depth. Water samplers are generally used from vessels, ships or fish trawlers. Bottle sample method is the simplest method which is generally used for the collection of water samples from any desired depth of shallow systems like the nearshore waters, estuaries and mangroves.

Meyer's Water Sampler

It consists of an ordinary glass bottle of about 1-2 l capacity and is enclosed with a metal band. It is weighted below with a lead weight and there are two strong nylon graduated ropes. One rope is tied to the neck of the bottle and the other to the cork. During operation, the corked up (closed bottle) is sent to the desired depth where, the stopper is jerked open by a strong pull of the cork rope. During this process, water flows into the bottles and then the cork rope is released to keep the cork closed. Afterwards, the bottle containing the water sample is taken out of the water column using the neck rope. This type of water samplers may, however, be used up to a depth zone of only 20m.

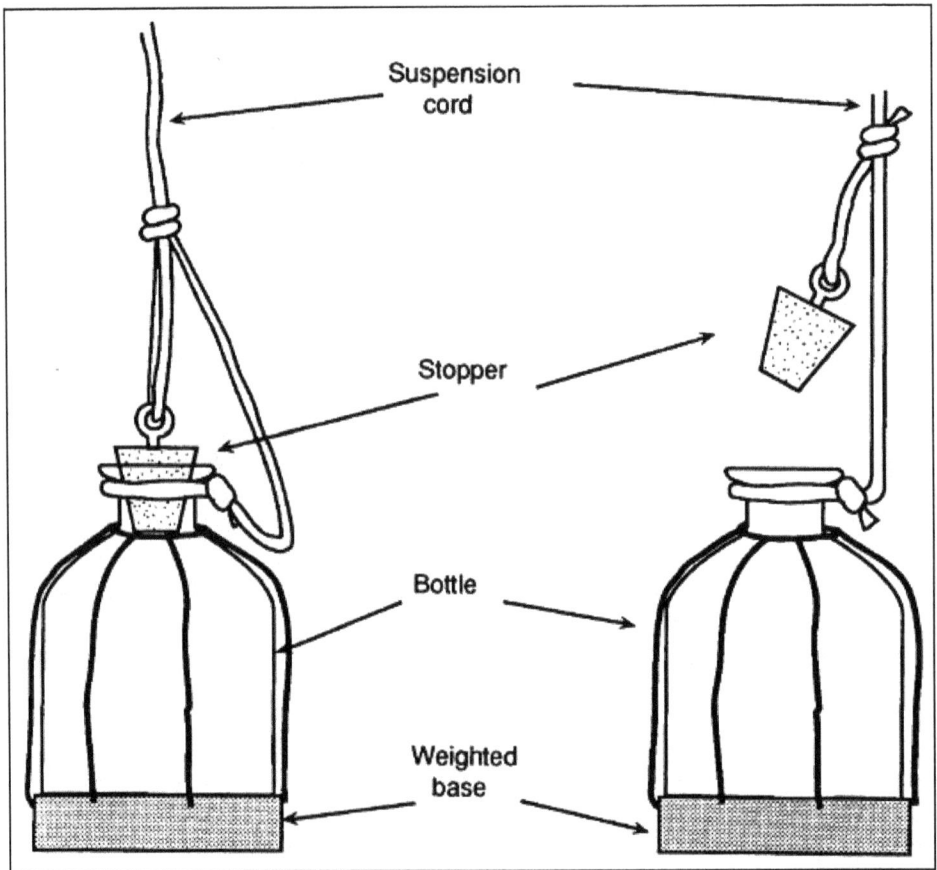

Friedinger's Water Sampler

It is made of Plexiglas or Perspex with two hinged covers. During operation, the sampler is sent down in an open state to the desired depth and is closed by a drop weight messenger, which falls down inside on sliding rail, closes the covers and makes the bottle water tight. By this way, the water with the planktonic organisms of the specified depth is trapped inside.

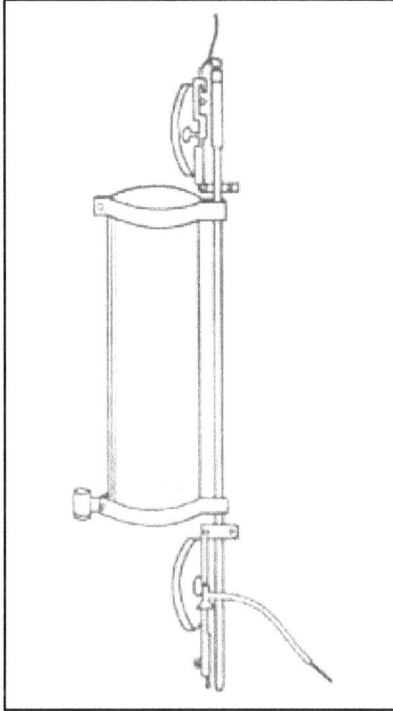

Nansen Reversing Water Sampler

Reversing water samplers available in different capacities (750-1250ml) are also used for taking water samples from sub-surface levels. Each sampler which is made of brass and plated inside with tin is fitted with a drain cock and an air-vent to facilitate draining the trapped water sample. It is provided with two plug valves, one on each end of the metal cylinder and is operated synchronously by means of a connecting rod fastened to the clamp which secures the sampler to the wire rope. When the sampler is lowered, the clamp at the lower end and the plug valves are in open condition, so that, the water of the concerned depth can pass through the sampler. The sampler is held in this position by the release mechanism which passes around the wire rope. When the messenger is dropped down the rope, it strikes the release and the sampler immediately falls over and turns through 180 degrees, shutting the valves which are then held closed by a locking device. For collection of water samples from different depths simultaneously, a series of water samplers are suspended one above the other from a wire rope and are lowered into differing depths in the open

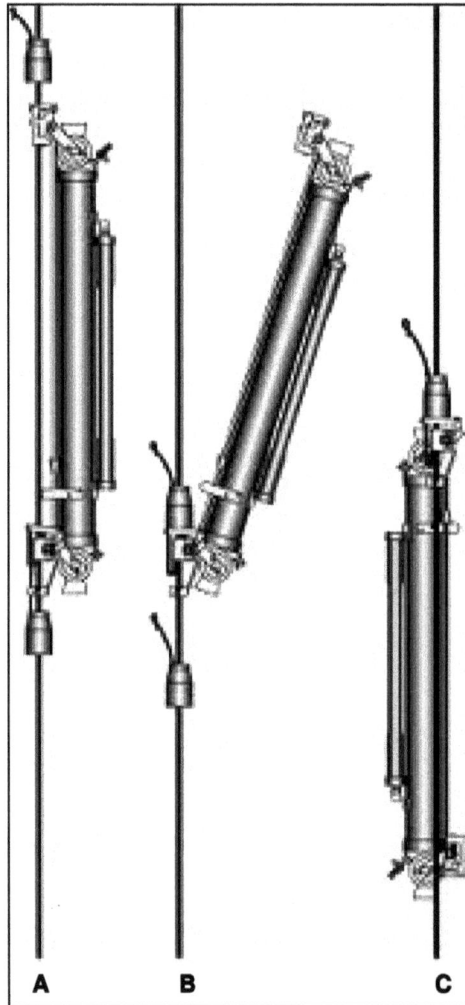

A. The Bottle is sent in open condition;
B.A Messenger is sent; bottle reverses and water of the depth fills and bottle closes;
C. Water filled /closed bottle is taken to the ship.

state. In this case, after reversing one water bottle, the messenger releases another messenger that is attached to the wire clamp before lowering. The second messenger closes the next lower sampler releasing the third messenger and so on.

Plankton Pumps

Plankton pumps, pump a continuous stream of water to the surface and the plankton sample can be rapidly concentrated by continuous filtration. As the pumps can collect continuously as the tube is lowered through the water column, the samplers are integrated from surface to the desired depth.

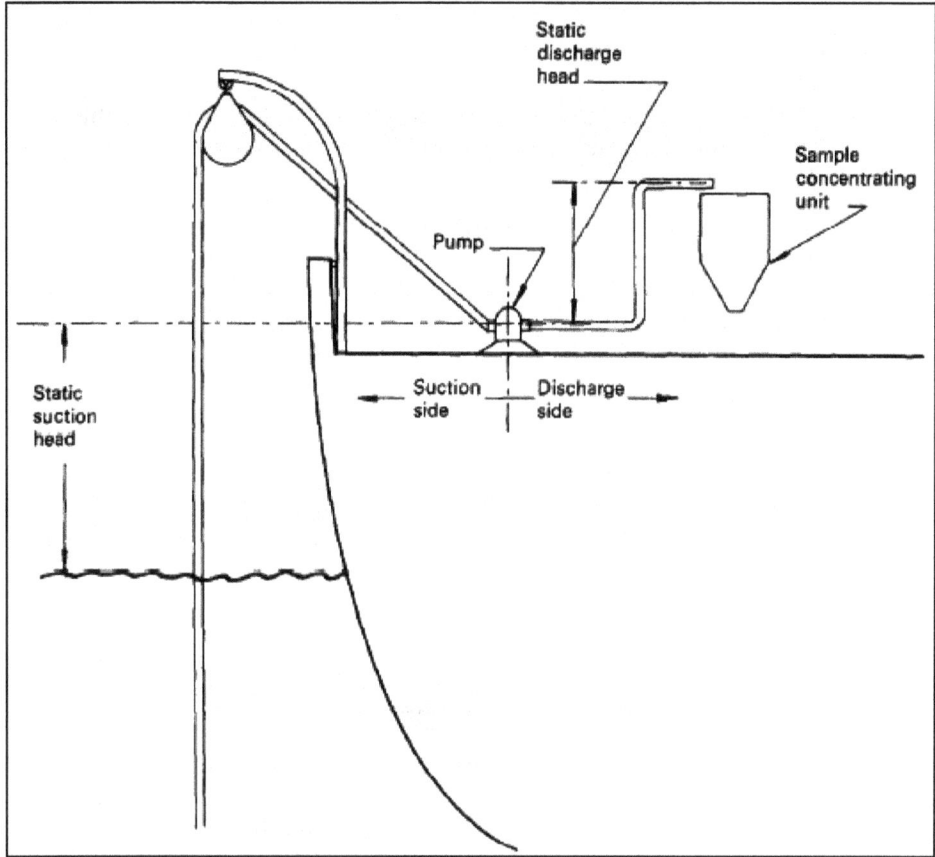

2.2. Collection of Microplankton

The most common method of microplankton (net plankton) collection is by a net. The amount of water filtered is more and the net is suitable both for qualitative and quantitative studies. The plankton nets used are of various sizes and types. The different nets can broadly be classified into two categories, the first being the open nets which are used mainly for horizontal and oblique hauls and the second being the closed nets which are with messengers for collecting vertical samples from desired depths. The vertical haul is made to sample the water column. The closing net is lowered to the desired depth and hauled slowly upwards. The plankton sample collected is from the water column traversed by the net. Closing mechanisms are used to sample a specific body of water. The samples taken with closing nets are analyzed to study plankton species composition and abundance at different depths.

In the horizontal sampling, the net is towed at a slow speed usually for 5 to 10 minutes. The towing speed of the net should be such that the maximum amount of water enters through the mouth of the net for better filtration and net used can withstand the strain. The towing speed of the net recommended for horizontal sampling is 1.5 to 2.0 knots. Most of the plankters are known to migrate vertically in response to light conditions. Hence, their occurrence is normally poor in the upper layers during daytime. For better quantitative and qualitative plankton collections, the suitable time for horizontal plankton sampling would be before dawn, after dusk or night. The horizontal collections are mostly carried out from the surface and subsurface layers.

Observed Values of Mesh Widths of Various Monofilament Nylon Nets

Net No.	Av.mesh width (um)
0	508 (Coarse net)
3	306
4	288
8	184
12	130
14	101
16	88
18	72
20	65
25	55
30	41 (Fine net)

The netting of the filtering cone is made of bolting silk, nylon or other synthetic materials. There is a great variety of mesh available from the finest to the coarse pore sizes. Interestingly, as the number of net increases, the mesh width decreases. The nets with finer mesh will capture smaller organisms, larval stages and eggs of planktonic forms and fish eggs while those with coarse netting material are used for collecting bigger plankton and fish larvae. In addition to the mesh size, the type, length and mouth area of the net, towing speed, time of collection and type of haul will determine the quality and quantity of plankton collected. The mesh size of the netting material will influence the type of plankton collected by a net.

Typical Plankton Net

A typical plankton net usable in the surface layers is conical in shape and has the following structure. A net ring made up of stainless steel and wrapped and sealed with polythene tubing is present anteriorly. To this, a non- filtering portion made of a coarse khaki cloth is attached using button and hole system. The filtering portion is made of monofilament nylon material with suitable mesh size and is followed by again a non-filtering portion of khaki cloth. To the latter, a metal net bucket is provided with a stop cock which is tied with a strong twine. Alternatively, to the net bucket, a nylon net piece of same mesh size (used in the filtering portion) may also be tied with a strong twine.

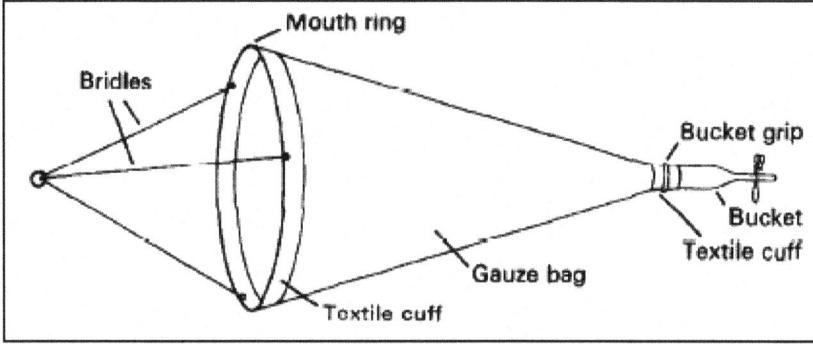

Plankton Net in Operation

Student Net and Wisconsin Net

The two most common types of plankton nets are the "Student" net and the Wisconsin net. Both nets are generally used to strain a large volume of water (usually more than 200 L) and reduce it to about 20 ml. The inside of the net must be washed from the outside prior to emptying the jar (Student net) or bucket (Wisconsin net). This rinses into the funnel or bucket any organisms that may be clinging to the net mesh.

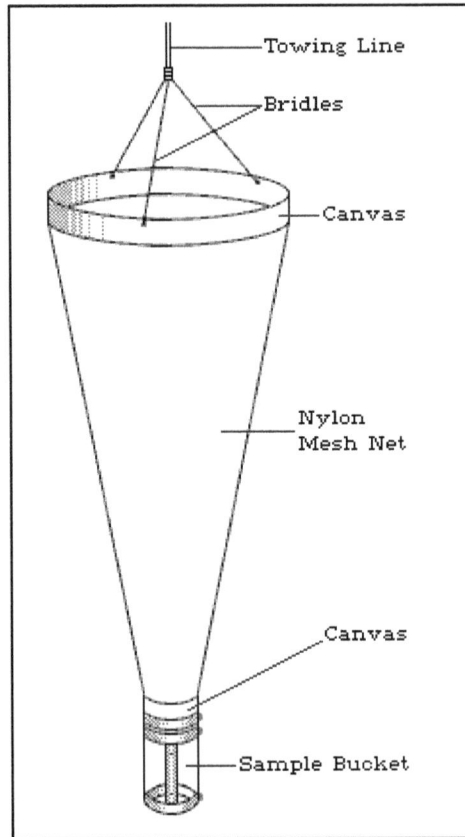

Student Net

Wisconsin Net

Numerous types of the simple, non-opening/closing plankton nets are available. Most are simple ring-nets with mouth openings ranging from 25 to 113cm in diameter and conical or cylinder-cone nets of 300-500cm in length. Among the ring-nets that have been widely used is the Indian Ocean Standard net. Earlier nets were made from silk, but today nets are made from square mesh nylon netting. Typical meshes of nets used for sampling plankton range from 60-200 microns. Most of these nets are designed to be hauled vertically. They are lowered to depth cod-end first and then pulled back to the surface with plankton being caught on the way up.

Indian Ocean Standard Net

This net has a mouth area of one square meter and a total length of 5 meters. The net is made of nylon gauze with a mesh size of 330um.

Indian Ocean Standard Net

Closing Nets

Typical Closing Net

The closing net resembles the normal plankton net in many respects and differs from the latter mainly in the presence of a closing mechanism. Due to the opening and closing mechanism of this net, the collection of plankton between the surface and bottom or from any desired depth in the vertical profile is possible. The bag of the net in its lower part has a fine filtering portion, above which, is a strip of coarse khaki cloth and between this and net ring, a coarse net work is sewn. To the net ring is attached three suspension ropes all tied to a small brass ring. This brass ring is further attached above the closing mechanism to the two ropes, *viz.* the tow and the closing ropes. The net is sent down with a weight at the bottom and is pulled up by the tow rope. The closing rope is lead quite loosely round the part of the net made of coarse cloth by means of rings sewn in. One end of it is firmly tied to one of the rings and the other end is attached to the release mechanism. When the messenger is sent through the tow rope, it strikes the release mechanism and the net is then freed from the tow rope and rests slowly on the closing rope which fastens like a ring round the bag of the net when it is drawn up and this closes the net. After the operation, the net should be drawn out immediately. As no more plankton enters the closed net during the taking up, the plankton caught belong to the depth hauled only.

Clarke-Bumpus Horizontal Closing Net

Of the various instruments which have been devised for quantitative collection of plankton, the Clarke-Bumpus plankton sampler is the best one.

This instrument is simply a plankton net connected to a flow meter which allows measurement of the volume of water filtered by the sampler. It can be handled easily from a small craft, and can be equipped with nets of different mesh sizes to collect plankton samples. It can be utilized advantageously for the collection of plankton

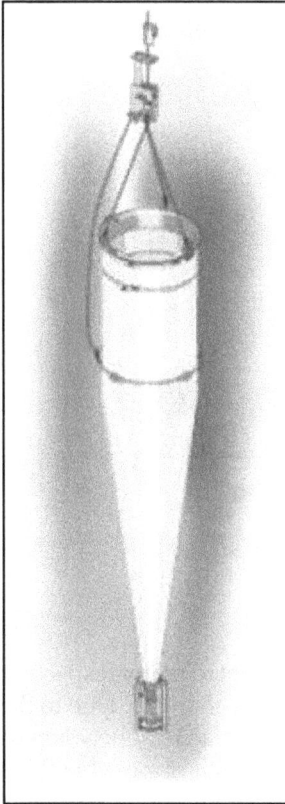

Closing Net–Open Condition

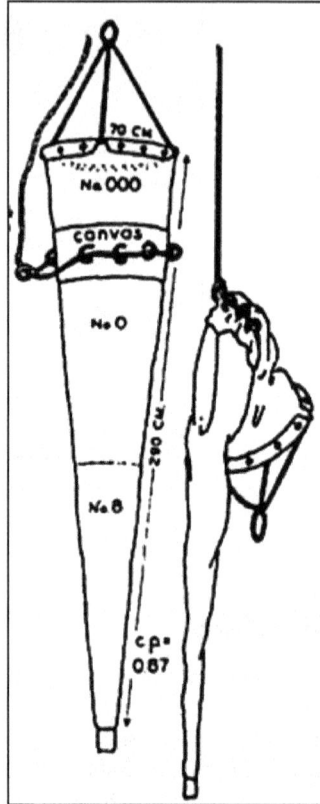

Closing Net–Closed Condition

samples along vertical, horizontal, or sinusoidal hauls within a selected layer of water. When it is used for vertical hauls, it has the advantage over simple, ordinary plankton nets of directly measuring the volume of water filtered, and consequently, has greater accuracy. Its most effective use, however, is for horizontal or sinusoidal hauls. This sampler is equipped with a mechanism for opening and closing the mouth distance by means of messengers. This means that even very deep layers can be sampled singularly. In addition, a retaining device for a second pair of messengers makes possible the use and the simultaneous functioning of more than one sampler at different depths using a single cable. At present, there are two kinds of Clarke-Bumpus samplers in the market, one with a mouth opening of 7-12 cm and the other with a mouth opening of 5-30 cm.

Hardy's Continuous Plankton Recorder

This sampler is torpedo shaped and rapid plankton samplings are possible with this instrument, besides collecting data on the distribution of species over a large distance within a short time. It consists of a water tunnel, two rolls of net silk

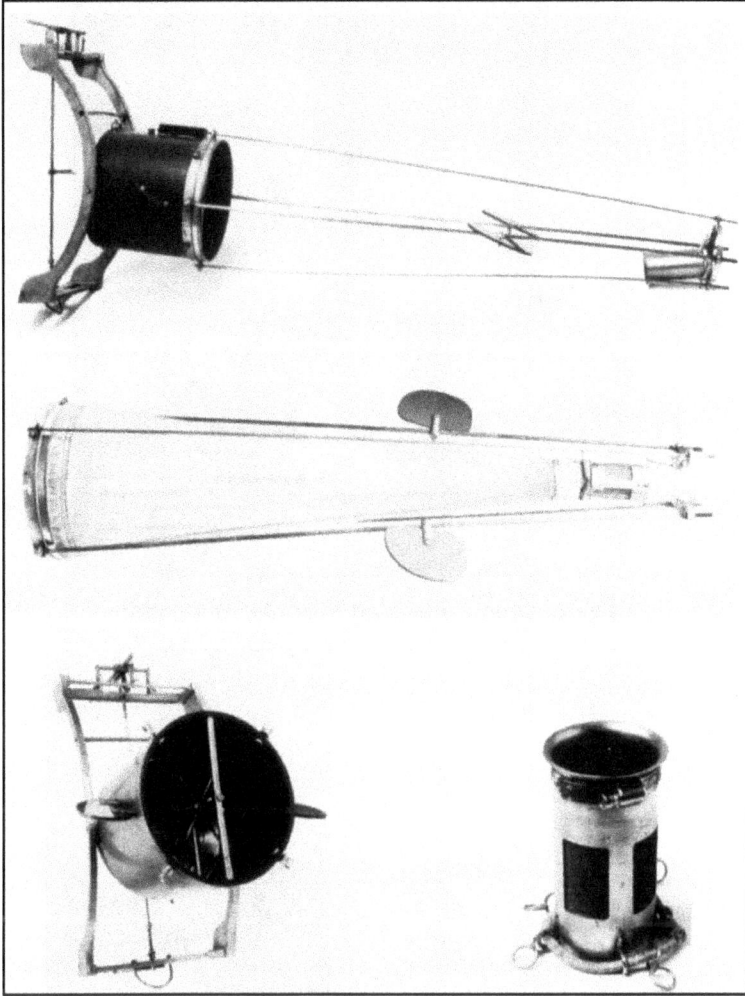

**Clarke-Bumpus Horizontal Closing Net with Filtering Portion,
Flow Meter and Net Bucket.**

(24 meshes/cm; mesh width, 270um) of about 15cm width, gear mechanism, propeller, stabilizing fin, horizontal stabilizing fin, diving plane, vibration damper, storage spool, formalin tank, etc. When the instrument is towed from a vessel, the propeller is turned by the passage of the water. Following this, the gears and the adjoining spools containing net silk strips are simultaneously activated. When the water with plankton flows through the tunnel, one of the rolls of silk from the lower spool runs up through the water tunnel in the same way as a film in a camera and filters the plankton which streams through the small squarish mouth (1.27 cm square) of the instrument. Before this net strip containing filtered plankton reaches the spool which is immersed in the formalin tank, another roll of silk from the upper spool which acts as a covering strip covers the collecting strip and keeps the plankton on it. Then both the strips are

Hardy's Continuous Plankton Recorder: In Operation.

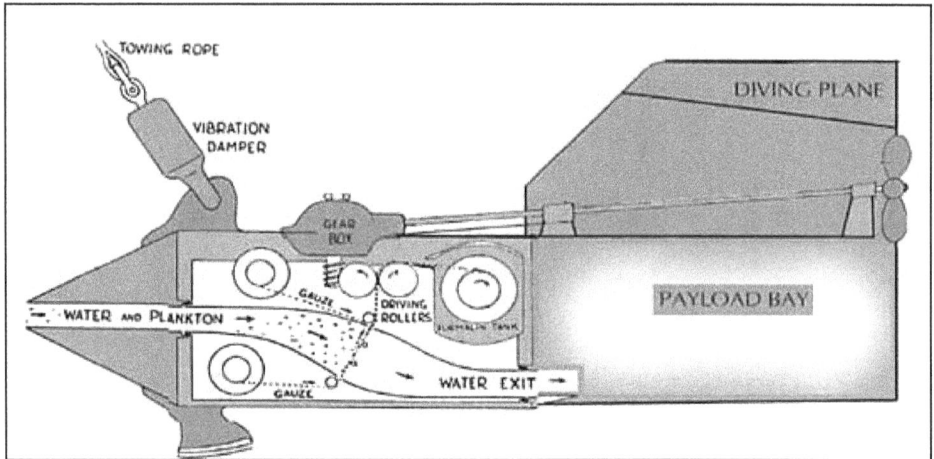

Hardy's Continuous Plankton Recorder: A Cutaway View.

wound up on the storage spool which is located in a container filled with formalin solution to preserve the plankton in the strips. When the storage spool is full, the same is removed and sent to a laboratory for analysis.

Chapter 3
Plankton Processing Procedures

3.1. Fixation and Preservation of Phytoplankton

It is very important to fix and preserve the phytoplankton using suitable fixatives and preservatives as soon as they are collected, to prevent the adverse effects of light and temperature which might cause rapid decay of organisms. A very widely used fixative and preservative for plankton is formalin. The commercial formalin is obtained as 40 per cent (saturation limit) formaldehyde dissolved in water. The formalin has to be stored in inert glass or plastic containers and not in metal containers as the formalin reacts with the latter. The commercial formalin may also contain dissolved impurities such as iron and formic acid, which disintegrate the shells of some planktonic organisms. The acid content of commercial formalin however, may be neutralized, by the addition of excess of calcium carbonates.

Neutralized Formaldehyde

It is made of 1 litre of 20 per cent Formaldehyde solution (HCHO) + 100g Hexamethylene tetramine. It is a general preservative for all phytoplankton samples. For water samples containing nanophytoplankton, 2 ml of this fixing/preserving agent is added for every 100 ml of water sample. For net samples, this fixing/preserving agent is added to make up about onethird of the volume if the samples are dense.

Acidified Formaldehyde

This agent is prepared by mixing 20 per cent Formaldehyde solution (HCHO) with 50 per cent Glacial acetic acid (CH$_3$COOH). It is a good preservative for all phytoplankton especially diatoms, but not for coccolithophorids, as the acid may dissolve coccoliths. A very widely used fixative and preservative for plankton is formalin. The commercial formalin is obtained as 40 per cent (saturation limit) formaldehyde dissolved in water. The formalin has to be stored in inert glass or plastic containers and not in metal containers as the formalin reacts with the latter. The commercial formalin may also contain dissolved impurities such as iron and formic acid, which disintegrate the shells of some planktonic organisms. The acid content of commercial formalin however, may be neutralized, by the addition of excess of calcium carbonates. For preserving net and other phytoplankton sample, 2 per cent neutralized formaldehyde (*i.e.* formalin) may be used.

Lugol's Solution

It is a good preservative, especially for flagellated and ciliated phytoplankton to retain the flagella and cilia. It consists of 10g iodine and 20g potassium iodine dissolved in 200ml of distilled water and 20g of glacial acetic acid. The solution can be made up a few days ahead and stored in a dark bottle for convenience. Lugol's solution is added in a ratio of 1 part to 100 parts of the seawater sample. For about 250ml of water sample containing nanophytoplankton, five drops of this preservative is quite sufficient. However, for phytoplankton samples containing more of coccolithophorids, this preservative should not be used as these cells may turn black with this preservative. Further, brownish coloration of the phytoplankton caused by Lugol's solution poses a problem in taxonomic investigations. Such unwanted coloration may be removed by oxidizing the Lugol's solution using a few drops of 3 per cent sodium thiosulphate solution.

Osmic Acid

This preservative is prepared by adding 200 mg of osmium tetroxide in 10 ml of distilled water. This preservative is added at the rate of 3-6 drops per 100 ml of phytoplankton sample.

Glutaraldehyde Solution

This preservative is prepared by mixing 8g of glutaraldehyde in 100ml of distilled water. This solution is added to the phytoplankton sample in the ration of 1:1. The phytoplankton samples containing more of diatoms should not be stored in glass bottles owing to their silica content.

3.2. Zooplankton Specimen Processing Procedures

Zooplankton meant for the taxonomic study need to be narcotised, fixed and preserved in that order, immediately after they are caught. Otherwise, autolysis, bacterial action, cannibalism or chemical deterioration may set in.

Narcotisation

The initial reactions of zooplankters to any fixative and preservative are rapid and jerky movements, contraction of body and appendages. This can hinder species identification subsequently. This is overcome by temporarily anesthetizing the specimens with suitable narcotics (analgesics). It has been found that full necrosis occurs in about 24hrs in most of the zooplankters.

The narcotising solutions recommended are carbonated water (1:20 by volume), chloroform(1ml), methyl alcohol (50 per cent) and magnesium chloride (7 per cent). The carbonated water is usually used as it is cheap and easy to use in the field. The collected specimens should not be transferred directly to narcotising solution. The narcotising fluid is added drop by drop to the water containing the specimens. The specimens are not to be kept there for a long period to avoid any damage. As soon as the morphological characters are observed, the specimens are washed with distilled water and put back into the fixative. It is always better if the sample is narcotised within 1 hour of sample collection.

Fixation

The narcotised sample should be fixed within 5 minutes. Formaldehyde is a suitable fixative for general taxonomic work and for bulk material. Several formulations are available with formaldehyde.

Formaldehyde

The most common method is to dilute commercial formaldehyde (40 per cent), so that it makes either 2 per cent or 4 per cent with seawater. Four per cent formaldehyde is prepared using 10ml of 40 percent commercial formaldehyde (borax-buffered) and 90 ml of seawater. On the other hand, 2 percent formaldehyde is prepared using 5 ml of 40 percent formaldehyde (borax-buffered) and 95 ml seawater. Before diluting the strong, commercial formaldehyde, 2 g of borax is to be added to every 98 ml of 40 percent formaldehyde. This will raise the pH to around 8 to 8.2. If a lower pH is needed, sodium glycerophosphate is used at the rate of 4 g to 96 ml of 40 per cent formaldehyde.

Additives which bring anti-oxidant properties are known to improve the sample resistance to bacteria and moulds. These additives which give flexibility to tissues may be employed with formaldehyde. A concentrated solution of these additives with formaldehyde is as follows:

Propylene phenoxetol (50 ml) is first mixed with propylene glycol (450 ml) and this mixture is added to 500ml of 40 per cent commercial formaldehyde (buffered with borax). 10 ml of this strong solution with 90 ml of seawater will produce a good, mild fixative suitable for most plankters.

The jar after fixative is added is to be gently rotated so that, the contents are well mixed. The jar is also inverted several times at 10 min intervals during the first hour of fixation.

Clearing

The fixed specimens should be cleared of any attached material such as detritus or precipitate. This can be done by removing the extraneous substances with fine forceps/needles without damaging the specimens. The specimens are immersed in clearing fluids such as lactic acid, glycerin or propylene glycol. The lactic acid is commonly used as a clearing agent and care should be taken that the specimens are not left in the lactic acid for a long time, as this would result in the disintegration of the body tissues of zooplankton. Examination of external features becomes easier after clearing the specimens. To study the internal structures, staining of specimen is required.

Preservation

After 7-10 days, the fixative may be replaced by the preservatives. In order to replace the fixative with the preservative, the fixative is to be decanted carefully, taking care not to lose any part of the sample. The sample jar is then filled about three-quarters with seawater or distilled water. To this, the correct volume of concentrated preservative (2.5 ml of buffered 40 per cent formaldehyde) is added and the jar is filled completely with seawater or distilled water and seal it with a secure lid. The jar is to be inverted several times to scatter the animals through the fluid. Once the preservative has been added and the sample stored, the container is checked for signs of evaporation, tightness of the lid and animal condition at the end of 1 month, 6 months and then yearly.

The other preservative used is 70 per cent ethanol or 40 per cent isopropanol. The ethanol is used for preserving museum specimens but it is costly and volatile. Glycerin is often added to formalin to prevent shrinkage of specimens, drying of the material and to facilitate retaining the colours of zooplankters. For better shelf life of the zooplankton samples, the preservative should be changed within the first 6 months.

It would be better to store the preserved zooplankton samples in a well ventilated room at a temperature less than 25°C. The samples should be kept in the wide mouth glass jars. A good quality pre-printed labels, in which the collector's name, fixative and preservative used and other field information are written should be put into the jars for ready reference at the time of sample analysis.

3.3. Bottling and Labelling

Bottling

Storage of plankton samples, especially diatoms in bottles made of soft glass is preferred. Plastic bottles are not suitable for storage of diatoms with delicate frustules or spines. Further use of glass of very low quality for storage of diatoms may result in precipitates. The bottles are closed with a leak proof cork. After the analysis of the plankton content of the bottles, wax coating is given around the cork of the bottle after the latter's closure for permanent storage of plankton. This would also help in avoiding the loss of formalin by evaporation in the long run.

Labelling

Proper labelling of the collected and bottled plankton samples is essential. All types of information regarding plankton collection should be written on the labels, so that the plankton samples can also be identified later accurately. The label should contain enough information about the sample collected in order to assure proper identification of the sample. The label is written with light colored water-proof marker or wax pencil.

Chapter 4
Staining and Mounting Techniques

4.1. Mounting of Phytoplankton

Preparation of Phytoplankton Specimens for Light Microscopy

Common diatoms can be identified by examination of raw (without acid cleaned) material in a water mount. Common diatoms such as *Chaetoceros* spp. and *Rhizosolenia* spp. are identified by their gross morphology, and special structures like *Chaetoceros* setae and the shape of the *Rhizosolenia* its and process. However, this method is not effective for identifying the essential morphological structures of other genera; for example, the areolation and processes of *Coscinodiscus* and *Thalassiosira,* and the striation and raphe structure of *Navicula* and *Pesudonitzschia*. Organic part and cell contents, which obscure the image, have to be removed. Acid cleaning is one of the methods used to separate diatoms frustules into single valves on which structures of diatoms are best seen.

Acid Cleaning

Before acid cleaning, salt particles associated with the diatoms in a test tube should be washed by rinsing and centrifuging in distilled water. Then, the test tube with the sample is allowed to dry by removing water. Adding some hydrochloric acid to the test tube dissolves the calcareous matter, and also loosens any diatoms that may be attached to the debris. After allowing the test tube with the sample for one or two days, the test tube is well shaken and the solid matter, including the diatoms is allowed to settle at the bottom. The acid is then decanted off and the sediment is washed by adding water, and pouring off again after allowing time for the solids to settle. Finally, most of the water is poured off and concentrated sulphuric acid is added slowly and carefully. Until red fumes are no longer evolved, small crystals of potassium dichromate are then added at intervals. The sulphuric-chromic acid mixture is then poured off and water is added. Acid and dichromate treatment must be repeated until cleaning is complete, if the diatoms are not yet properly cleaned with water.

Mounting

For mounting, diatoms are put in a drop of distilled water on a cover slip that has been smeared with a little Mayer's egg albumen, which is prepared by mixing 50ml of white of egg with 50ml of glycerin and 1g of sodium salicylate. After allowing the water to evaporate, the diatoms on the coverslip are thoroughly dried by heating, and then using any mounting media like Canada balsam, Styrax, Hyrax or DPX mount can be done. After cooling the specimen-mounted slide excess resin is trimmed off by a knife, and the preparation is finally sealed with nail polish or wax. Glycerin mounting and polyvinyl lactophenol mounting are other methods of mounting diatoms. These are more convenient to mount the diatoms in slides directly by embedding them in polyvinyl lactophenol. Canada balsam is ideal for permanent mounts. For longer preservation, diatoms can also be cleaned and stained with methylene blue and Bengal pink. Subsequently, they are embedded in Canada balsam in microscopic slides and covered with cover glasses.

4.2. Staining, Dissection and Mounting of Zooplankton

Staining and Dissection

Staining of specimens is required to study the internal structures of any zooplankton specimen. Light staining of the specimens is carried out by adding a few drops of rose Bengal, lignin pink, chlorazol black E and methylene blue added to the lactic acid. Borax carmine is used for staining small zooplankton, larval stages of crustaceans and icthyoplankton (fish eggs and fish larvae). The lignin pink and chlorazol black E penetrates the chitin and stains the internal tissues and facilitates dissection. Rose Bengal also facilitates dissection. It is usually added to the zooplankton samples when the preservation is done. The dissection of stained specimens is carried out under stereoscopic dissecting microscope with fine needles on the cavity slides. Two dissecting needles should be used, with one needle the specimen is held firmly and with the other body somites are cut. One should be careful while dissecting the delicate mouthparts. The dissected mouthparts and other structures are immersed in glycerin or lactic acid before putting in the mounting medium.

Mounting

Permanent glass slides are made by using the natural or synthetic resins. Canada balsam, gum chloral, glycerin jelly and lactophenol are normally used as mounting agents. The Canada balsam dissolved in xylene or benzene is used for whole mounts. The disadvantage with balsam is that the mounts may become dark with time. Lactophenol is widely used and this can be stored for a long time. Before mounting, the whole specimen or the dissected parts, the slides and coverslips are thoroughly cleaned with ethanol and dried. A few drops of mountants are placed on the glass slide and then the specimens or their dissected structures are transferred. The coverslip is supported by fragments of broken cover slip or wax. The slides should be completely dried and stored in a slide box for subsequent examination for species identification.

Chapter 5
Estimation of Biomass (Standing Stock)

The term standing stock of phytoplankton represents the total number of cells present in a unit volume of water. The zooplankton standing stock or biomass denotes the live weight or the amount of living matter present in a unit volume of water. In the case of zooplankton, prior to determination of biomass, larger zooplankters such as medusae, ctenophores, salps, siphonophores and fish larvae should be separated from the zooplankton sample and their biomass is taken separately. The total biomass of zooplankton would be the biomass of bigger forms, plus the biomass of the rest of the zooplankton.

5.1. Volumetric Methods

Displacement Volume

The total plankton volume is determined by the displacement volume method. In this method the plankton sample is filtered through a piece of clean, dried netting material. The mesh size of netting material should be the same or smaller than the mesh size of the net used for collecting the samples. The interstitial water between the organisms is removed with the blotting paper. The filtered plankton is then transferred with a spatula to a measuring cylinder with a known volume of 4 per cent buffered formalin. The displacement volume is obtained by recording the volume of fixative in the measuring jar displaced by the plankton.

Sedimentation Volume

By this method, the net plankton samples are transferred to a graduated cone *i.e.* sedimentation chamber and are allowed to settle for about a day. The volume (ml) of the samples is then recorded directly. The actual volume of plankton is then calculated per cubic metre of water filtered.

Sedimentation Chambers

There are several drawbacks in the above volume methods. In these methods, the total volume of both phytoplankton and zooplankton is determined. Hence, determination of volume of either phytoplankton or zooplankton is not possible using the above methods.

$$\text{Volume of zooplankton (ml/m}^3) = \frac{\text{Total volume of zooplankton}}{\text{Volume of water filtered (m}^3)}$$

5.2. Gravimetric Methods (Wet Weight and Dry Weight)

Wet Weight

The gravimetric methods are more suitable for the estimation of zooplankton biomass only. The estimation of wet weight of zooplankton is carried out by filtering the zooplankton through a bolting silk cloth piece of less than 20um mesh size. The interstitial water is usually removed by blotting paper. The zooplankton weight is taken on predetermined or weighed filter paper or aluminum foil. The wet weight of zooplankton is expressed in grams.

Dry Weight

The dry weight method is dependable as the values indicate the organic content of the plankton. Analysis such as the dry weight is determined by drying an aliquot (subsample) of the zooplankton sample in an electric oven at a constant temperature of 60°C. The whole or total sample should not be dried because the subsequent analysis, such as enumeration of common taxa and identification of their species would not be possible after drying the sample. The dried aliquot is kept in a desiccator until weighing. The values are expressed in milligram. As the collected microplankton (net plankton) sample constitutes both phytoplankton and zooplankton, the above gravimetric methods may not give a true picture of phytoplankton or zooplankton biomass.

$$\text{Wet weight of zooplankton (g/m}^3) = \frac{\text{Total wet weight of zooplankton}}{\text{Volume of water filtered (m}^3)}$$

$$\text{Dry weight of zooplankton (mg/m}^3) = \frac{\text{Total dry weight of zooplankton}}{\text{Volume of water filtered (m}^3)}$$

5.3. Chemical Method for the Estimation of Zooplankton Biomass

In this method, the live zooplankton samples are dry frozen. Before analysis, the samples are rinsed with distillated water. Measurements of constituent elements such as carbon, nitrogen, phosphorus and biochemical elements *viz.* protein, lipid and carbohydrates are made. Sometimes the biochemical values of a particular taxon and species are undertaken to evaluate food energy transfer at higher trophic levels. The calorific content of the plankton can be used as an index of zooplankton biomass. As the collected microplankton (net plankton) sample constitutes both phytoplankton and zooplankton, the above chemical method may not give a true picture of phytoplankton or zooplankton biomass.

5.4. Quantitative Analysis (Faunal Enumeration)

Information on the faunal composition and the relative abundance of different plankton taxa and their species is obtained by counting the plankters present in the collected samples. The enumeration of specimens in the total sample is laborious, time consuming and mostly impractical. The number of common plankton groups and their species observed in the samples may vary from tens to thousands. For enumeration, it is recommended that the subsample or an aliquot is taken for the common taxa. However, for the rare groups, the total counts of the specimens in the samples should be made. For enumeration of plankton, the subsample or aliquot of 10 per cent to 25 per cent is usually examined. However, the percentage of aliquot can be increased or decreased depending on the abundance of plankton in the sample.

Subsample (Aliquot)

Instruments are available for splitting the sample into the fractions. These are generally made of plastic with internal partitions.

Folsom Plankton Splitter

Using this instrument, the collected plankton sample to be sub sampled is poured into the drum and the drum is rotated slowly back and forth. Internal partitions

divide the samples into equal fractions. The fraction may be poured again into the drum for further splitting. The process is repeated until a suitable subsample is obtained for counting. The splitter is thoroughly rinsed to recover the organisms. The sample is usually split into 4 subsamples. One of the subsamples is used for estimation of dry weight, the second for counting the specimens of common taxa, the third for relative abundance of species and the fourth fraction is kept as reference collection. Plastic or glass pipettes are also used to take the subsample for counting. The Stempel pipette is used to obtain a certain volume (0.1 to 10 ml). The plankton sample in a glass container is diluted to a known volume and is stirred gently. The Stempel pipette is then used to remove the subsample or aliquot for counting.

Counting

Assessment of standing crop of plankton in different periods is essential for marine environment, as the level of biomass indicates directly or indirectly its fertility and fishery resources. The biomass may be estimated in various ways.

Inverted-Microscope Method (Utermohl Method)

This method is more suitable for estimation of the nanoplankton content in the seawater samples. It involves a combined plate chamber (compound chambers) consisting of a top sedimentation cylinder (10, 50 or 100 ml capacity) and a bottom-plate chamber. It is made of rectangular perspex plate, a ring and a circular base plate of cover slip thickness. This plate is designed in such a way to fit into the mechanical stage of the inverted microscope. The combined chamber is made ready for use by placing a sedimentation cylinder of desired capacity on the top of plate chamber. Seawater sample (containing nanoplankton) preserved with lugol solution is poured into the combined chamber to overflow and a square top plate is placed in position to eliminate dust and evaporation. After sedimentation (sedimentation time in hours is about three times the height of the sedimentation cylinder in centimetres), the sedimentation cylinder is slowly pushed away from the bottom-plate chamber by using the square top plate of the plate chamber.

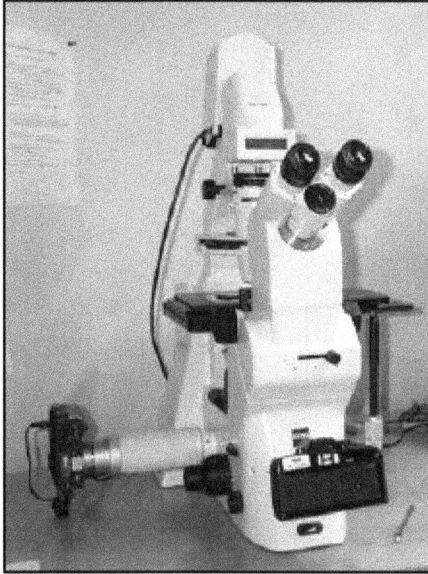

To enumerate the plankton, the entire bottom can be scanned at low magnification in the Utermohls inverted microscope.

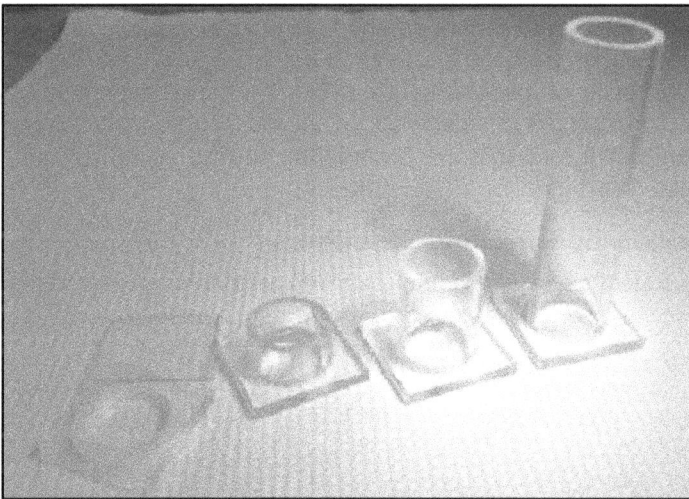

Utermohl Tubular Chambers

Calculation

The total number of organisms per millilitre is calculated using factors relevant to the volume sedimented, including concentration or dilution factors. The concentration of organisms C for each taxon is calculated according to: (a) for a concentrated sample, C [organisms/ml] = organisms counted/concentration factor; or (b) for a diluted sample, C [organisms/ml] = organisms counted x concentration

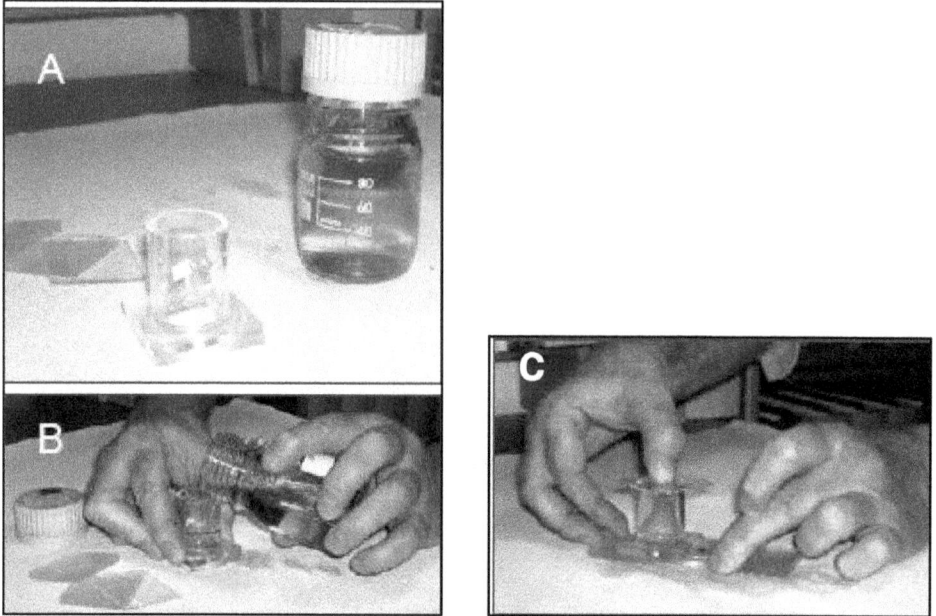

Stages in the Filling Up of Utermhl Tubular Chamber

factor. For example, after sedimentation, 100 ml of the original sample is reduced to 10 ml (concentration factor of 10), and a 1 ml subsample is taken for enumeration. In this, 600 organisms of species A are counted. The concentration of species A in the original water sample is calculated according to: C = 600/10 = 60 organisms/ml. To obtain a total density of organisms per millilitre, sum all counting results of individual taxa expressed as organisms per millilitre.

Enumeration by Counting Chambers

The enumeration of net plankton may also be done by various counting chambers, however, the most commonly used counting chamber is Sedgwick Rafter cell.

Sedgwick Rafter Cell

This counting cell is filled with the plankton sample with the help of a Stempel pipette and is placed on the mechanical stage of the microscope.

Stempel Pipette

Then the counting cell is left for about half-a-hour for proper sedimentation. The organisms are then counted from one corner of the counting cell to the other. The

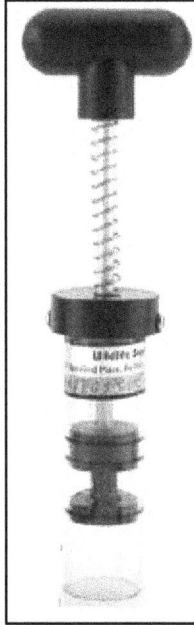

Sedgwick Rafter is moved horizontally along the first row of squares and the planktonic organisms in each square of the row are thus counted. The rafter is moved to the second row and organisms in each square here are counted. (Few transects may also be counted instead of all the squares). The total number of organisms is then computed by multiplying the number of individuals counted in transects with the ratio of the whole chamber area, to the area of the counted transects. Replication of counts of one ml samples is recommended for the statistical treatments. The average values are taken into account for calculation.

The total number of planktonic organisms present in a litre of water sample can be calculated using the formula:

$$N = \frac{n \times v}{V} \times 1000$$

where,

N = Total number of organisms per litre of water filtered;

n = Average number of organisms in 1 ml of plankton sample.

v = Volume of plankton concentrate (ml)

V = Volume of total water filtered (l).

Total number of organisms/m^3 = Number of organisms /l x 1000

Determination of Volume of Water Filtered by Circular Net

Ordinary Flow Meter

If the diameter of the net mouth is equal to that of attached flow meter, then the volume of water filtered= pie r2 h, where r2 is the radius (0.1m) of flow meter and h is the distance (100m) travelled by the net

$$V = \frac{22}{7} \times 0.1 \times 0.1 \times 100 = 3.11 \text{ m}^3 \text{ (or 3110 litres)}$$

Digital Flow Meter (model 4381100)

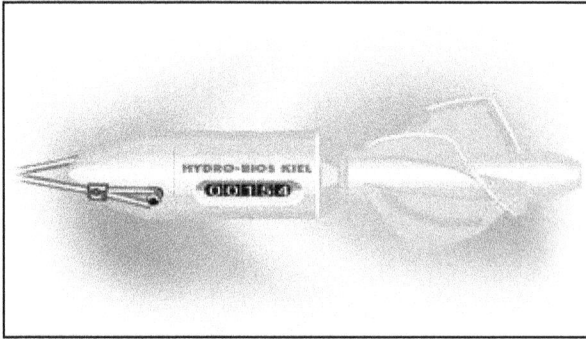

The plankton net model 438 030 has a diameter of 40cm *i.e.* the opening area is 0.125m2.If the number of revolutions associated with a tow is 266 (noted from the flow meter counter), the water volume passed through the plankton net is 266x0.3x0.125x1000=9,975 litres.

Where 0.3 is the pitch of the impeller *i.e.* 0.3m/revolution and 1000 denotes litres for 1 m3.

Determination of Volume of Water Filtered by a Square Net

$$V = S^2 d$$

where,

S = Length of the side of the frame.

d = Depth of haul.

Chapter 6
Estimation of Chlorophylls

This method is chiefly employed to estimate phytoplankton biomass. This is a rapid method for determining phytoplankton density in a sample involves the extraction and measurement of chlorophyll concentrations. Photosynthetic pigments include chlorophylls *a*, *b* and *c*, carotenoid pigments, and phaeopigments. Phaeophytins are simply the degradation products of chlorophyll molecules. Each of these chlorophyll pigments has a specific absorption coefficient, and by the measurement of absorption at different wavelengths, the relative contributions of each pigment to the total chlorophyll absorption can be determined. The most commonly used solvent for extraction of photosynthetic pigments has been acetone.

After the collection of the water sample, it is filtered through a Millipore (Pore size 0.45μm). The addition of 1 ml of a 1 per cent Magnesium carbonate suspension onto the filter paper to form a thin bed, would serve as a precaution against the development of any acidity and subsequent degradation of pigment in the extract. All steps should be carried out in the dark to avoid pigment breakdown. The filter containing the sample is placed in 90 per cent acetone in a plastic vial covered with aluminium foil and shaken vigorously and gently ground with a homogeniser to ensure dissolving of the filter (Millipore) before storage in the refrigerator for 20-24 hrs. Some recommend, the addition of 1 ml of a 1 per cent Magnesium carbonate suspension onto the filter paper to form a thin bed, which will serve as a precaution against the development of any acidity and subsequent degradation of pigment in the extract.

After 20-24 hrs of extraction in the cold and dark, the plastic vial containing filter paper is brought to room temperature and the volume brought up to the original level by the addition of 90 per cent acetone in a graduated centrifuge tube. The solution is centrifuged for about 20 minutes at 5000 rpm and the supernatant solution is considered for the determination of optical density, or transmission percentage which is mainly with the aid of a Spectrophotometer.

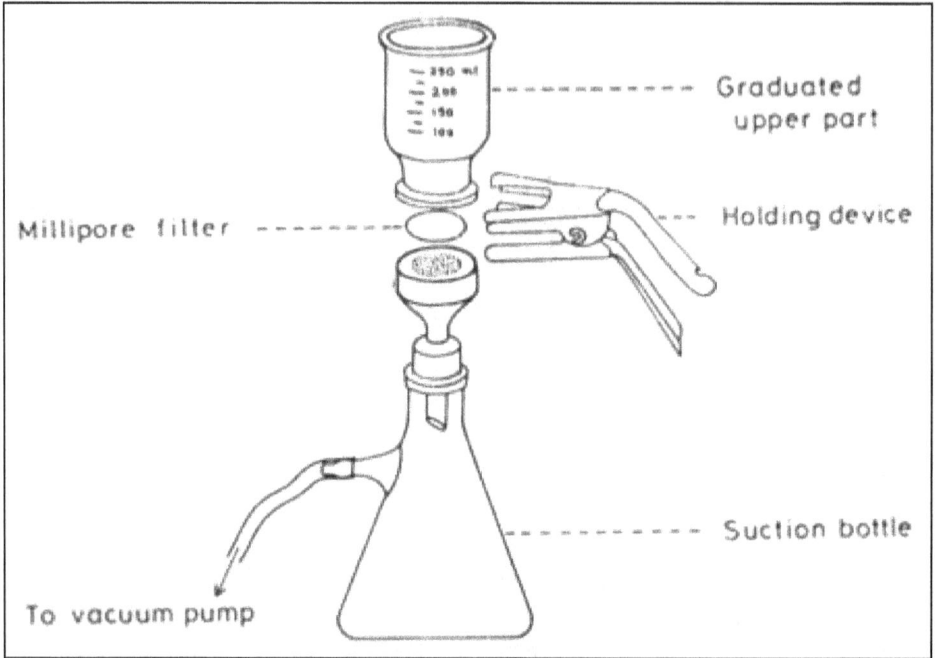

Millipore Filtration Unit.

6.1. Determination of Chlorophyll *a*, *b* and *c* in Spectrophotometer

The supernatant solution obtained above is transferred to a 1cm (path length) cuvette (absorption cell) of the Spectrophotometer for the determination of optical densities at different wavelengths *viz.* 664, 647 and 630nm, the maximum absorption wavelengths of chlorophylls *a*, *b* and *c* respectively. All the extinction values are corrected for a small turbidity blank by subtracting the 750nm value from the 664, 647 and 630nm absorptions.The individual chlorophyll contents of the water samples are determined using the following formulas:

Chlorophyll *a* =11.85E 664—1.54E647—0.08E630

Chlorophyll *b* =21.03E647—5.43E664—2.66E630

Chlorophyll *c* =24.52E630—1.67E664—7.60E647

where,

E is the absorbance at different wavelengths in the respective wavelength

Calculation

$$\mu g \text{ chlorophyll } a, b \text{ or } c / l = \frac{C \times v}{V \times 1}$$

where,

Ca, Cb and Cc are the three chlorophylls which are substituted for C in the above equation; v, volume of acetone (ml) used; V, volume of water (litres) filtered for the extraction of chlorophyll and 1, pathlength (cm) of cuvette used in the Spectrophotometer.

6.2. Estimation of Chlorophyll *a* and Phaeophytin *a*

The amounts of active chlorophyll *a* and phaeophytin *a* may also be considered as a measure of phytoplankton standing crop. For assessing the active chlorophyll *a* content of the sample, the optical density values of the solution at 750 and 665nm are measured in a Spectrophotometer. Subsequently, one drop of 3N hydrochloric acid is added to the above solution and the extinction values at 750 and 665nm are measured again for the estimation of phaeophytin a content.

Calculation

$$\mu g \text{ active chlorophyll } a/l = \frac{26.7\,[(665a\text{-}750a) - (665b\text{-}750b)]\,v}{V \times l}$$

$$\mu g \text{ phaeophytin } a/l = \frac{26.7\,[1.7(665b\text{-}750b) - (665a - 750a)]\,v}{V \times l}$$

where,

665a and 750a are the extinction at 665 and 750nm before acidification, 665b and 750 b are the extinction at 665 and 750 nm after acidification

v is the volume of extract (ml)

V is the volume of water filtered (l) and

l is the length of the lightpath (cm) or cuvette

6.3. Estimation of Plant Carotenoids

The estimation of carotenoids is done by measuring the sample prepared above at wavelengths 510 and 480nm.

Calculation

Plant carotenoids= 7.6 (E 480—1.49E 510)

where,

E is the absorbance at 490 and 510nm (corrected for 750nm) absorbance

Chapter 7
Production of Plankton

7.1. Estimation of Primary Production

Light and Dark Bottle Method

In marine ecosystems, estimates of the rate of gross primary production, net primary production and respiration rate can be calculated by measuring the rates of oxygen production and of oxygen consumption in a known volume of water using the Light and Dark Bottle Method. Samples of the collected seawater are placed in two glass bottles of the same volume. One bottle is left clear, and the other bottle is covered to exclude light. The initial O_2 concentrations are measured. Then the two bottles are returned to the light conditions from which they were taken for 24 hours. After 24 hours, the bottles are retrieved, and the final O_2 concentrations are measured.

Materials

- ☆ Two clear glass bottles of identical volume with lids or stoppers
- ☆ One light bottle and the other covered to exclude light (black painted -dark bottle)
- ☆ Aluminium foil
- ☆ Other sets of bottles are used to measure at different depths.
- ☆ Oxygen meter or glasswares and reagents for Winkler's titration
- ☆ A float and line or pole apparatus to hold the bottles in place at the depth where the samples are taken. A weight at the bottom of the line or lines will help to hold the bottles steady.

Procedure

Pre-lab Preparation (in the Lab)

Sample bottles are prepared. Each pair of bottles should consist of one clear (light bottle), and one covered to keep light out (dark bottle). The bottles need to be of identical volume, and should be labeled for the intended depth.

Light and Dark Bottles.　　　　**Light and Dark Bottles in Suspension.**

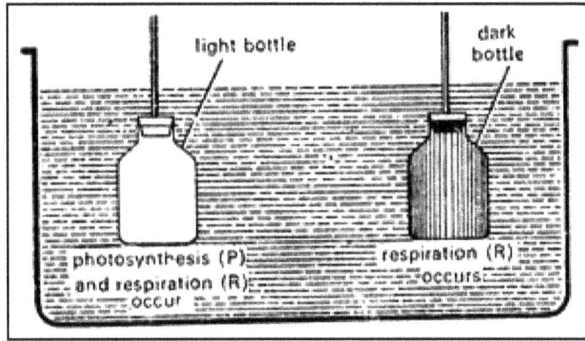

In the Field

Water samples are taken just below the surface by directly holding the sample bottle in the water. A third bottle is used to collect some water as before to top-off the sample bottles, if necessary, after O_2 readings are taken.

O_2 concentration of each bottle is measured by an O_2 meter. This should be the same as the samples are identical initially. If water is lost in the measurement, the same is replaced from the third bottle of sample that is identical to the sample pair. The caps are secured and the sample bottles are suspended in the area where water samples were taken with the apparatus constructed. The date and time, and the initial O_2 concentration as mg / L (this is also parts per million ppm), are recorded. If experiments are to be done for more than one depth, this procedure is repeated for the other sample pairs.

The experiments are run to allow respiration and photosynthesis to proceed for 24 hours.The sample bottles are retrieved after the period of the experiment is over and the O_2 concentrations are measured in each bottle.

Calculations

The respiration rate (R) is calculated in terms of oxygen consumption as follows:

$$R = \frac{C_0 - C_D}{\Delta t}$$

where,

C0 is the initial concentration of O_2 (in mg/l), CD is the final O_2 concentration in the dark bottle (in mg/l), and Δt is the period of time over which respiration took place. If Δt is measured in days, as in the procedure given, then R will be mg O_2/l / day; if Δt were in hours, then R would be mg O_2/l /hr.

Respiration rate = _____ mg O_2/l/day

The gross primary production is calculated as photosynthetic productivity of oxygen (PG) in mg O_2/l /day, as follows:

$$P_G = \frac{C_L - C_D}{\Delta t}$$

where,

CL is the final concentration of oxygen in the light bottle.

Gross primary production= _____ mg O_2/l/day

The net primary production (PN) (expressed in mg O_2/litre/day) is calculated as:

$$P_N = P_G - R$$

Net production = _____ mg O_2/l/day

C14 Method

Labeled carbon is probably the most extensively used procedure for oceanic productivity studies. This method proves to essentially advantageous because it is relatively safe, weak β-emission (0.15 Mev) as well as its long half life (4700yr), so that storage yields no major problems.

Procedure

The activity per ml of the working solution needed for the different productivity experiments depends on the production rates expected, duration of incubation, bottle size, etc. Invariably, 0.2-1 ml of the working solution is used per bottle containing water sample.

Seawater samples for which production rates are to be determined are first collected from the specified depths and are transferred to the light and dark bottles kept in a dark box. Then, a known dose of the working solution is injected rapidly into the bottles with the help of a graduated hypodermic syringe having a needle not shorter than 5 cm. The bottles are then incubated for a known period of time by suspending them in the respective depths from where the water samples were taken for the experiment. After the incubation is over, the experimental bottles are removed from the concerned depths and are stored in a light-free case until the filtration of water samples is done. During filtration, aliquots of water samples for filtration are rapidly transferred into a suitable vacuum filtration apparatus on to a No. 2 membrane filter or Millipore filter of about 0.5 µ porosity. The vacuum is applied at about 0.5 atm which will help avoiding damaging the phytoplankton cells. The filtration is done invariably in a semidarkend area.

The filters, after their removal from the filtration apparatus are placed onto planchets which are then kept in a desiccator containing silica gel. Filters obtained from light and dark bottles are then subjected to counting in a Geiger-Muller counter. Under constant light source, the rate of production is obtained in mgC/m³/hr by the following formula:

Where, cpm, counts per minute; the total CO_2 is assumed to be constant in oceanic waters and the value is 90 mg CO_2/l; 1.06, a correction factor for the isotope discrimination effect and to be used as the 14C incorporation will be slow compared to 12C; 1000 to convert the value for m3; 12/4 to get the value of C from CO_2 as the molecular weight of CO_2, 44 and the atomic weight of C, 12.

Rate of production = (photosynthesis) (mgC/m^3/hr)

Net activity (cpm of light bottle-cpm of dark bottle) X Total CO_2 cpm added X 1.06 X 1000 X 12/44

Hrs of incubation

7.2. Estimation of Secondary Production

The rate of increase in the biomass of heterotrophs per unit time and area is called secondary productivity. Secondary production is a comprehensive population variable that incorporates density, biomass, individual growth rates, population death rates, and life spans. It has not only been used in ecosystem energy flow studies, but also as a wide variety of ecological questions. It is superior to density or biomass alone as a response variable because it is a measure of function rather than structure.

Secondary production can represent the formation of mass for an entire trophic level. While population biomass units are often presented as grams/m^2, the typical unit for secondary production incorporates time (*e.g.*, grams m^{-2} year^{-1}, grams m^{-2} week^{-1}). One tends to think of biomass as a structural (or static) variable and production as a functional variable because the latter measures an ecological process through time.

It is well known that not all food eaten by an individual is converted into new animal mass. We can consider a zooplankton species grazing on phytoplankton. In this case, only a fraction of the material ingested (I) is assimilated (A) from the digestive tract; the remainder passes out as faeces (F). Of the material assimilated, only a fraction contributes to the growth of an individual's mass or to reproduction — both of which ultimately represent production (P). Most of the rest is used for respiration (R). A small portion of the energy is lost in excretion, but the latter is usually ignored in energy budgets. Simple equations are used to illustrate the fate of ingested energy, such as I = R + P + F. Alternatively, production is P = I - F - R.

Production is a growth process that must first add to the mass of individuals before it can be consumed in the process of flowing to a predator or the next trophic level. So it is a two-step process: individual growth first creates new population biomass, then some of the biomass flows to the next level as a portion of the individuals consumed by predators.

Calculation of Secondary Production by Sampling a Cohort (Single Age Group)

After hatching occurs, the density of a cohort declines, and individuals increase in mass. For any interval of time between sampling dates, such as between 6 and 7, production is easily calculated as the product of the increase in mass (ΔW) and mean density between dates ($\tilde{N}$).

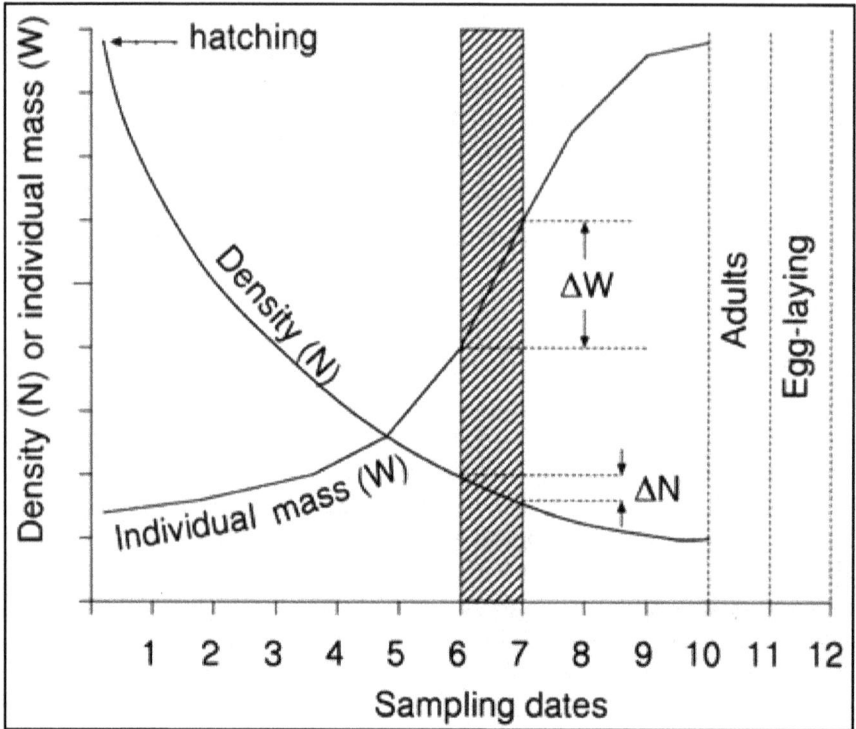

A Simple Method for Measuring Secondary Production.

The easiest production methods to understand are from field sampling in which a single age group (or cohort) is followed from birth/hatching through time as illustrated in the figure. Following hatching of this hypothetical population, only mortality (a decline in numbers) and individual growth (an increase in mass) occur through the rest of the life cycle of this generation.

If the population is quantitatively sampled at regular intervals, production between each interval is calculated from population density and individual mass. The shaded bar in the figure represents how the population changes between sampling intervals 6 and 7. There is an increase in individual mass (ΔW) and a decline in density (ΔN). Production during the interval is calculated as ΔW times the mean density ($\tilde{N}$) between dates. Total cohort production (P) is the sum of production for all time intervals:

$$P = \Sigma \tilde{N} \Delta W$$

This simple approach accounts for biomass losses due to mortality as well as the accumulation of biomass through individual growth. Production of all time intervals is added to obtain annual production.

Chapter 8

Micrometry and Size Measurements

By micrometry, while viewing through a microscope, the length, breadth and other details of an organism are measured. The size determination of the planktonic organisms forms an important aspect, especially, in preparing the report on the occurrence of new species or taxonomic studies for publication.

Total Magnification in Microscope

Ocular Lens	Objective Lens	Total Magnification
10x	4x	40x
10x	10x	100x
10x	40x	400x
10x	100x	1000x

8.1. Ocular and Stage Micrometers

Ocular Micrometer

In micrometry, an ocular micrometer (graticule) plays an important role. The ocular micrometer is a circular glass piece which contains a scale of lines, which are engraved or photographically reproduced. This scale is of 10 mm in length divided into ten equal divisions. Thus, on the scale of the ocular micrometer one hundred divisions of 100 μm each are present.

Stage Micrometer

Stage micrometer is a microscopic slide of 7.5 x 2.5 cm, on which, a scale of 1mm long is engraved, divided into 100 divisions of 10μm (0.01mm) each.

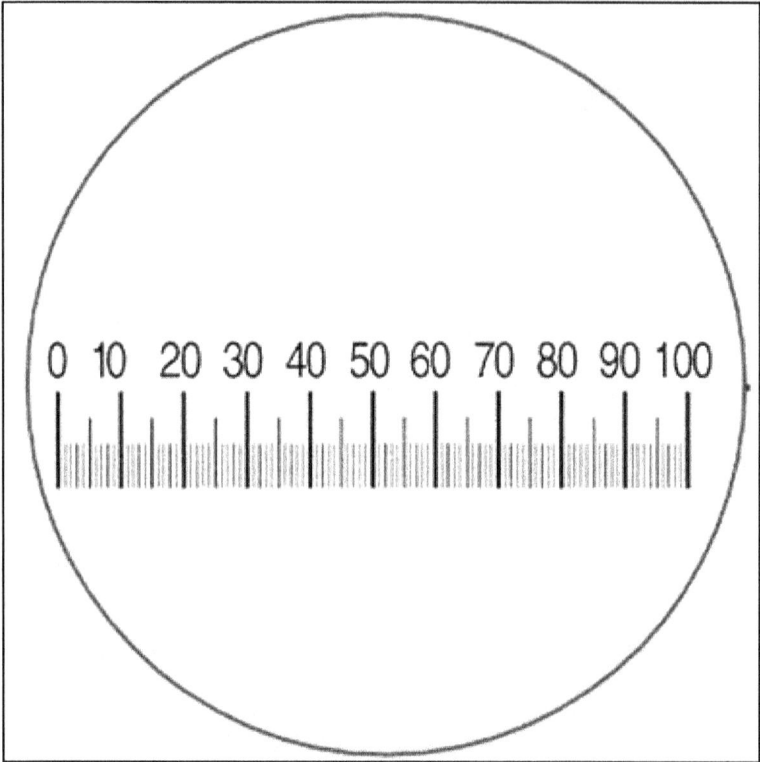

Ocular Micrometer

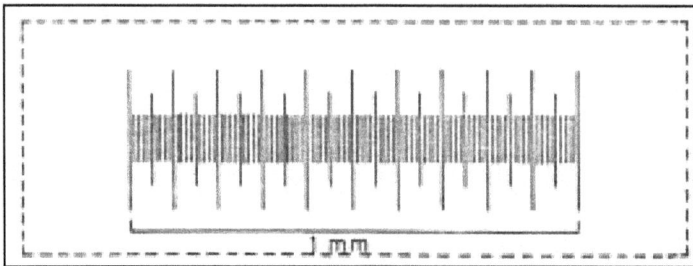

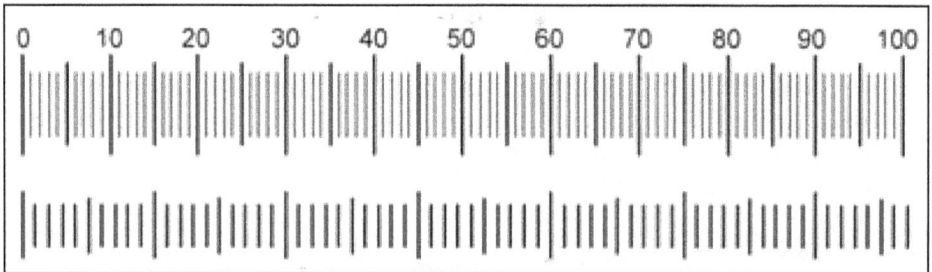

Graticule Divisions (Above) and Stage Micrometer Divisions (Below)

8.2. Calibration of Ocular Micrometer

Ocular micrometer is mounted on the diaphragm inside the eyepiece of the chosen microscope at the focal of the eye lens. On the diaphragm inside the eyepiece, at which point, the image from the object is also focused, so that, the two can be viewed simultaneously. Now, not only the object in focus, but superimposed on the object, the series of lines of the graticule is equally visible.

For the calibration of the graticule, the stage micrometer is first placed on the stage of the microscope. Then it is focused and aligned with the ocular micrometer scale. The stage micrometer is then moved carefully until its zero line is in exact coincidence with that of the ocular meter, in order to find out how many divisions on the ocular micrometer scale correspond with a certain number of divisions on the stage micrometer scale. From this, the value (in µm) of one division of the ocular micrometer under the chosen microscope with fixed objective and eyepiece powers is calculated.

If 40 divisions of the ocular micrometer correspond with 10 divisions of the stage micrometer scale, then these 40 divisions are equivalent to 100 µm. In other words,

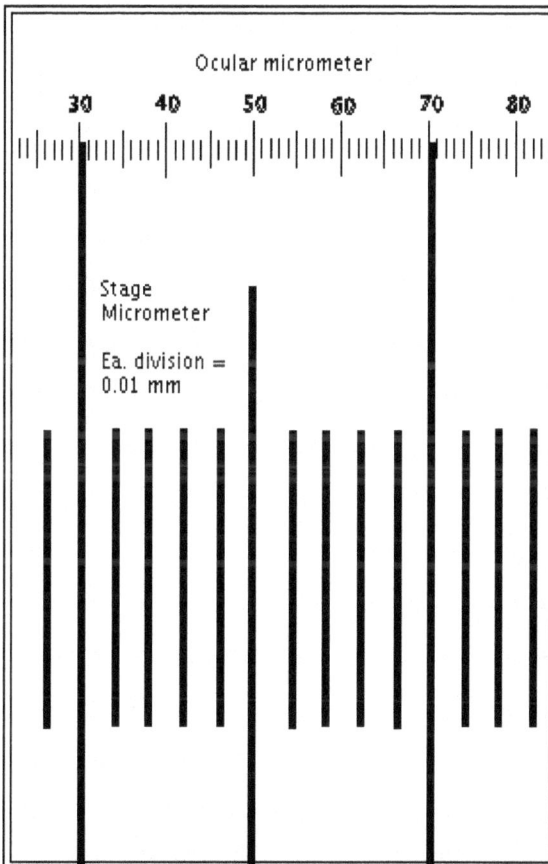

Calibration of Ocular Micrometer

these 40 divisions occupy 100 µm space of the stage micrometer (as one division occupies 10 µm of the space in the stage micrometer and the total length of the scale is 1000 µm –1 mm). Thus, one ocular micrometer division is equal to $100/40 = 2.5$ µm. This calibrated value of the ocular micrometer is of a particular objective and eyepiece of a microscope.

8.3. Size Determination

If the size of an object is to be determined in different objective or eye lens, the ocular micrometer scale is calibrated for all the combinations of the different objectives and eyepieces, all value may be tabulated and can be used whenever it is required.

The size of an individual species may be determined using the calibrated ocular micrometer and micrometer as follows. For size determination, on the stage of the microscope, the specimen for which the size is to be determined is now placed instead of the stage micrometer. If the length of an organism is to be determined, then the zero of the ocular micrometer is focused against the tip of the organism and the number of divisions of the ocular micrometer occupy the tail end of the organism, through which the length of the organism is found out. The number of calibrated ocular micrometer divisions multiplied by the corresponding calibrated value would give the total length of the said organism. For examples, if the graticule divisions are 60 then the total length is 60 x 2.5 µm= 150µm.

Chapter 9
Species Diversity

Species diversity is defined as the number of species present in an area. The species diversity of the plankton samples collected from different locations is calculated by the following methods.

9.1. Species Diversity Indices

Diversity index as a measure of species richness (Gleason 1922)

$$d = \frac{S-1}{\text{Log } eN}$$

where,

S, number of species in the sample; N, number of total individuals in the sample.

Margalef's Diversity Index (1968)

$$D = \frac{1}{N} \log 2 \frac{N!}{Na!\ Nb!.Ns!}$$

where,

Na! Nb! .... Ns!, the numbers of individuals of species a,b,....s; N, the total individuals. This index will be of more use, especially for the following changes in diversity through the dynamics of a mixed population.

Shannon - Wiener Diversity Index

Shannon-Wiener Index denoted by $H = -\text{SUM}[(pi) \times \ln(pi)]$

where,

SUM = summation

pi = proportion of total sample represented by species i

Divide no. of individuals of species i by total number of samples

S   = number of species, = species richness

Hmax   = ln(S) Maximum diversity possible

E   = Evenness = H/Hmax

Example 1: Uneven Zooplankton Community

Species (i)	No. in Sample	pi	In(pi)	(pi) x In(pi)
A. gracilis	60	0.60	-0.51	-0.31
T. discaudata	10	0.10	-2.30	-0.23
L. acuta	25	0.25	-1.39	-0.35
P. danae	1	0.01	-4.61	-0.05
M. norvegica	4	0.04	-3.22	-0.13
S = 5	Sum = 100			Sum = -1.07

H   = 1.07

Hmax   = ln(S) = ln(5) = 1.61

E   = 1.07/1.61 = 0.66

Example 2: Even Zooplankton Community

Species (i)	No. in Sample	pi	In(pi)	(pi) x In(pi)
A. gracilis	20	0.20	-1.61	-0.32
T. discaudata	20	0.20	-1.61	-0.32
L. acuta	20	0.20	-1.61	-0.32
P. danae	20	0.20	-1.61	-0.32
M. norvegica	20	0.20	-1.61	-0.32
S = 5	Sum = 100			Sum = -1.61

H   = 1.61

Hmax   = ln(5) = 1.61; H = 1.61/1.61 = 1.00

9.2. Similarity Index

Similarity index (S) is a simple measure of comparing experimental stations belonging to different biotopes for getting an integrated picture of the biotopes. It is calculated by the following equation.

$$S= \frac{2C}{a+b} \times 100$$

where,

C is the number of species at any two stations, a is the number of species at one station, and b is the number of species at the other station. In this index, calculation, all species carry equal weight irrespective of abundance or rareness.

9.3. Coefficient of Community

It is used to identify associations of species between any two stations (samples).

$$CC = \frac{c}{a+b-c} \times 100$$

where,

a is the number of taxa (species) in the first sample, b in the second sample and c common to both. CC relates to the likeness of samples on the basis of similar distributions of constituent species, without regard to aspects involving relative abundance. Lowest limits (values) indicate high affinity and one half of the lowest value is taken as separating set with low and intermediate affinity.

Chapter 10
Identification of Phytoplankton

Marine phytoplankton are microscopic and unicellular floating organisms, which are the primary producers supporting the pelagic food-chain. The two prominent groups of phytoplankton are diatoms (Bacillariophyceae) and dinoflagellates (Dinophyceae). Other minor groups of phytoplankton are the extremely small, motilealgae *viz.* naked microflagellates, silicoflagellates and cyanophytes (blue-green algae). A total of 888 species of diatoms, 211 species of dinoflagellates, 15 species of silicoflagellates, 15 species of coccolithophores and 81 species of cyanophytes have been reported from tropical seas.

10.1. Diatoms

Identification of diatoms in water samples is usually best done by using phase contrast optics, which reveal especially well lightly silicified structures, like delicate *Chaetoceros* setae, and also the organic chitan threads in Thalassiosiraceae. It is essential to know which side of the diatom cell is viewed. Intact single cells with a short pervalvar axis tend to lie up under the coverslip (*Coscinodiscus radiatus* and *Pleurosigma* sp). Diatoms like *Corethron* and *Rhizosolenia* with a pervalvar axis longer than the cell diameter or the apical axis turn girdle side upwards. Colony types like *Chaetoceros, Fragilariopsis* and *Thalassiosira*) are normally seen in girdle view in a water mount. Diatoms like *Thalassionema, Asterionellopsis* and *Pseudonitzschia* show either valve or girdle side. Cylindrical and discoid diatoms are readily recognized by the general circular outlines in valves view. When the cells are viewed properly the next step is to look for special features like setae in Chaetoceraceae, shape of linking processes in *Skelotonema* and in unpreserved material, organic threads from the valve in Thallassiosiraceae.

Frustular elements cleaned of organic material may also be oriented in various ways in a permanent mount. Flattened valves with a low mantle will usually be seen in valve view (Some *Coscinodiscus* spp., most *Navicula* spp.), while valves with a high mantle and protuberances may appear in girdle view (*Eucampia* and *Rhizosolenia*).

Lightly silicified bands shaped as those in *Rhizosolenia* and *Stephanopyxis* often lie with the girdle side up.

Structure of a Typical Diatom Cell

All diatom species are unicellular or colonial belonging to the Class Bacillariophyceae. They are pigmented and photosynthetic. Each cell is covered by a cell wall made of silica and takes the shape of a pill box with an overlapping lid called as frustule. The chloroplasts are invariably golden-brown. The diatoms are of two major groups, namely the centric diatoms (Order: Centrales) and the pennate diatoms (Order: Pennales) and are distinguished from each other on the basis of differences in cell wall structure and in the presence and absence of a raphe. The silica shell (frustule) of a pennate diatom is elongate and usually bilaterally symmetrical in valve view, with a lanceolate or elliptical outline. The frustule consists of two halves: the hypotheca (box like) and the epitheca (the overlapping lid- like in the box). The structure of the valve of a centric diatom is basically radially symmetrical, the frustules often resembling a Petri dish. As in pennate diatoms, the shells are ornamented with species-specific patterns and structures. In many centric diatom species, the valves contain radical rows of small, more-or-less hexagonal chambers, called 'loculate areolae'. The valve view with the pattern of sculpturing is very much clear in most of the species of *Coscinodiscus*. In other genera, this "pill box" shape is less apparent. For example, in the species of *Rhizosolenia,* the valve is often conical and the girdle length may be up to fifty times the valve diameter, and in *Chaetoceros,*the long spines, called setae, arise from the valves. Many centric diatoms also form chains of cells, in which the cells are connected together by all, or part of; their valve surfaces (*e.g. Chaetoceros, Lauderia* and *Eucampia*). In other genera, the cells are linked in chains by spines, *e.g. Skeletonema* or by mucilaginous threads arising from the valve surface, as in *Thalassiosira.*

The diatoms reproduce by binary fission. During division, new valve and girdle bands are formed. Thus, in each generation, one of the daughter cells is smaller than the parent cell. As there is considerable reduction in size after several divisions, the auxospores are formed to restore the original size of the species. When conditions are unsuitable for further cell division, the auxospore may develop into a resting stage.

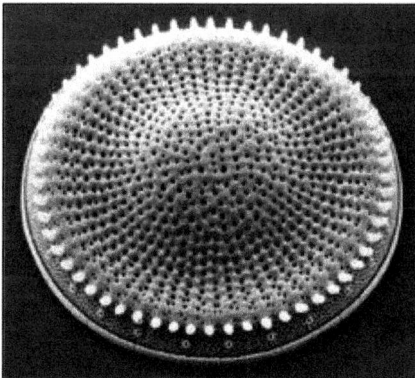

Radial Symmetry (Centric)   **Bilateral Symmetry (Pennate)**

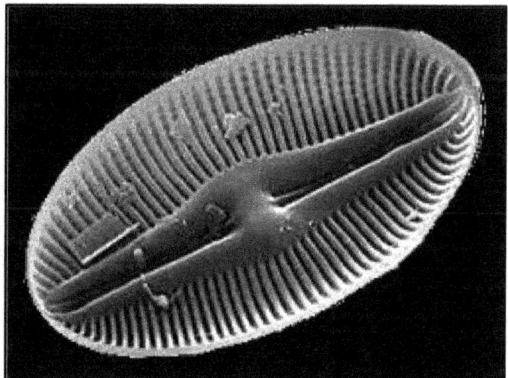

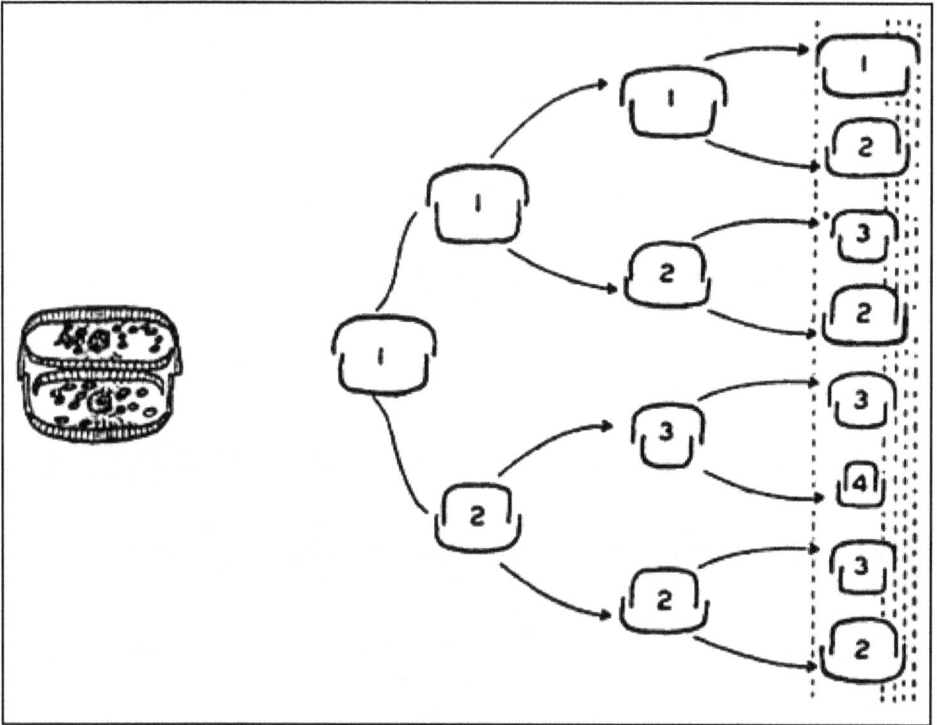

Binary Fission in Diatoms

The diatoms can be divided into two orders *viz.*, the Centrales and the Pennales. This classification is based largely on the symmetry and orientation of the secondary structures on the valve surface of the diatom species. Centric diatoms are non-motile, whereas many species of the pennate diatoms exhibit a gliding movement which is, in some way, dependent on the presence of a raphe.

Sexual reproduction of the Centrales is oogamous whereas that of the pennate species is generally isogamous. Although species of the two groups inhabit the same regions, Centrales are more commonly planktonic and marine, whereas Pennales are more dominant in estuarine and fresh waters.

The detailed classification of diatoms depends almost entirely on the structure of the siliceous skeleton. The Centrales are divided into three major groups, on the basis of cell shape and on the presence or absence of particular processes. For example genera such as *Coscinodiscus, Cyclotella* and *Melosira* are disc-shaped with no processes, whereas the valve surfaces of genera such as *Biddulphia* and *Chaetoceros* have various horns. On the other hand, genera such as *Rhizosolenia* and *Corethron* also have a complex girdle structure.

The classification of the pennate diatoms is largely based on the nature of development of the raphe. *Tabellaria* and *Asterionella* are examples of the group of diatoms which only possess a pseudoraphe.

The valves of *Eunotia* show the beginnings of raphe development. *Achnanthes* and *Cocconeis* have a raphe on one value only, but genera like *Navicula, Bacillaria* and *Nitzschia* have a raphe on each valve.

Class: Bacillariophyceae (Diatoms)

Order: Centrales

Thalassiosira eccentrica (Ehrenberg) Cleve (= *Coscinodiscus eccentricus*)

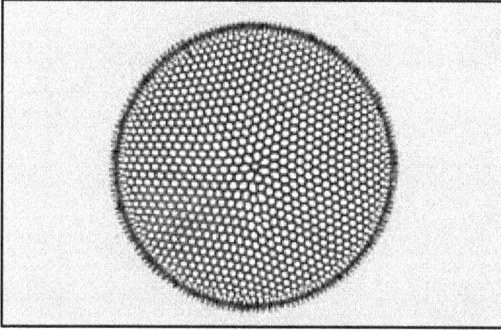

Description: Cells are solitary or forming chains joined by mucilage threads. Valves are disc-shaped and flat, and rounded. The Pervalvar axis is about one-sixth of the length of the cell diameter.

Dimension: Valve diameter, 25-70 µm.

Cyclotella meneghiniana Kutzing

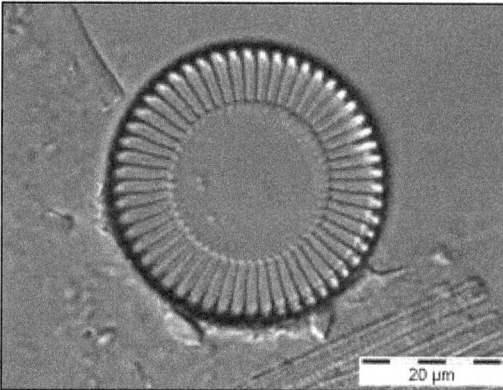

Description: Cell disc shaped with a number of regularly arranged striations which do not reach the center;

Dimension: Valve diameter, 17-24 µm.

Cyclotella striata (Kutzing) Grunow

Description: Cells are with evenly striated border; central area of the cell coarsely punctuated;

Dimension: Valve diameter, 15-30 µm.

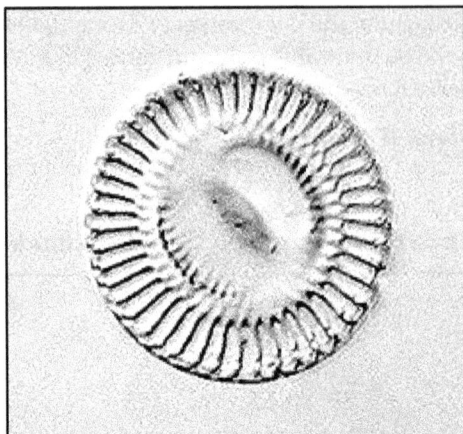

Asterompalus flabellatus (Brebisson) Greville

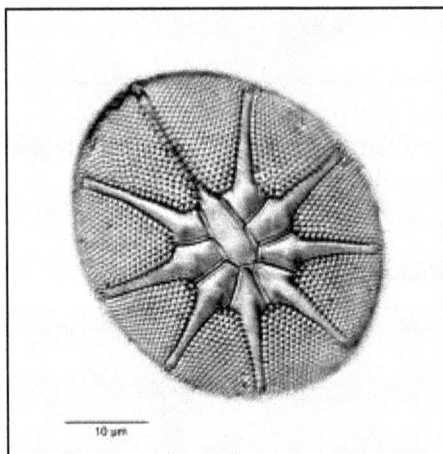

Description: Cell slightly convex; valves slightly ovate; middle sector lines unbranched; 7-8 slightly curved hyaline rays, of which one is narrower.

Dimension:Valve length 37-63 µm and breadth 32-55 µm.

Asterompalus wyvillei Castracane

Description: Valves rounded with 15 straight hyaline rays, of which, one is narrower; sector lines branched; dia., 71-74 µm flattened at ends numerous disc-shaped chromatophores.

Dimension: Valve diameter, 3-16 µm and length 10-91 µm.

Planktoniella sol (Wallich) Schütt

Description: Cells are discoid, solitary, with a central body surrounded by a wing-like expansion. Central or valvar portion small, valves convex covered with large polygonal areolation. Girdle with a continuous wing.

Dimensions: Diameter of the valve portion 30-360 µm.

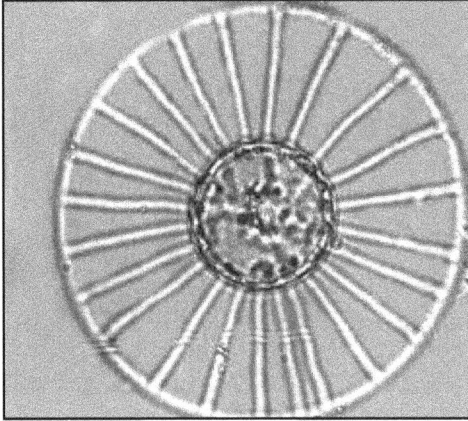

Lauderia annulata Cleve

Synonym: Lauderia borealis

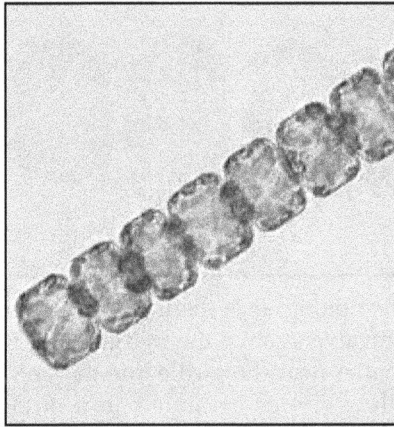

Description: Cells shortly cylindrical, linked to form straight filaments. In girdle view, the well-rounded valves give a beaded appearance to the filament. The Pervalvar axis slightly longer than diameter. Plastids, many, discoid, lobed. Valves very delicate, with fine radial costae radiating from a central annulus.

Dimension: Diameter of valve 30-50 µm.

Skeletonema costatum (Greville) Cleve

Synonyms: Stephanopyxis costata

Description: Cells discoid, oblong or weakly spherical. Valves mostly convex, but sometimes flat. Cells united to form filaments by a marginal ring of long spines; spines straight, filaments straight, sometimes slightly spiral. The spaces between the individual cells are frequently larger than the cells themselves due to the length of the spines.

Dimensions: Diameter 8-20 µm.

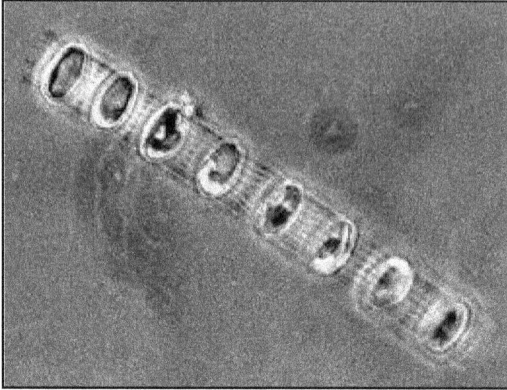

Stephanopyxis palmeriana (Greville) Grunow

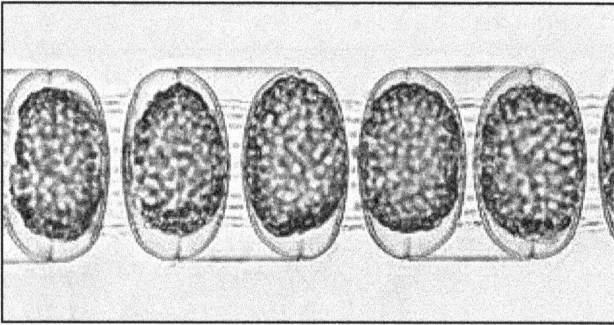

Description: Cells oblong. It is distinguished by the slight narrowing of the cylindrical part of the valve against the margin, and by the hexagonal areolation which are slightly smaller near the girdle line than on the areolae. Cells united in chains by 10-22 hollow spines arranged in a circle at each end of the cell. Chromatophores: numerous, plateline. Nucleus central.

Dimensions: Diameter of cells 27-71 µm.

Coscinodiscus jonesianus (Greville) Ostenfeld

Description: In girdle view, cells are as high as wide. Valves convex, slightly concave in the center. In valve view, central rosette of larger areolae more or less distinct. Areolae in radial and spiraling rows. Inside the marginal ring, two larger processes with prominent external areolated protuberances.

Dimensions: Diameter 140-250 µm.

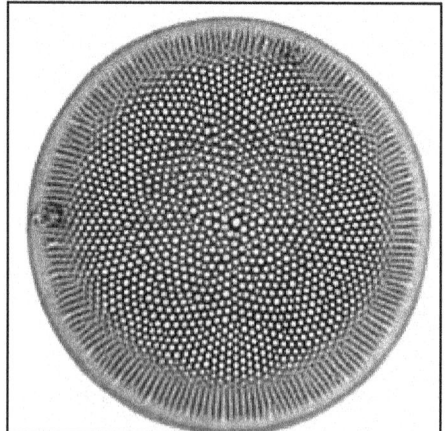

Coscinodiscus oculus-iridis Ehrenberg

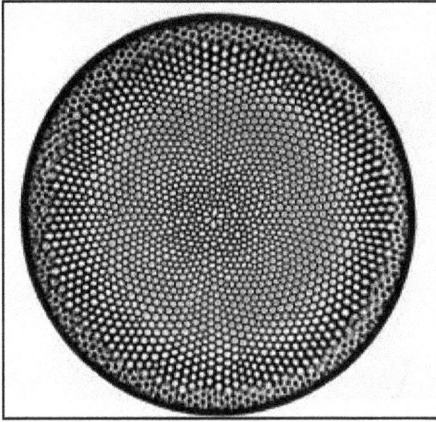

Description: Cells discoid, solitary, large. Valves almost flat, or slightly convex. Valves covered with large polygonal areolation, arranged in radiating lines, the lines long and short. Central rosette often large, consisting usually of five areolae, but sometimes fewer. Areolae small at the center of the valve, increasing gradually in size as they proceed to the periphery. Chromatophores: numerous rounded bodies. Areolae form an irregular cluster at the center of the valve, much less pronounced than the other species, the radiating striae have small areolae at the center.

Dimensions: 192um.

Hemidiscus cuneiformis Wallich

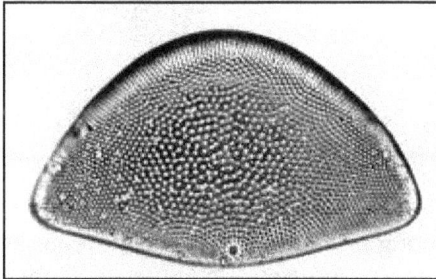

Description: Ventral margin regularly and gently convex. Dorsal margin strongly convex, in some individuals remarkably high-arched, without constrictions near the valve poles. Apical axis along straight edge, 100 μm long; transapical axis 85 μm long.

Dimensions: Length of valve 80-120 μm, breadth 10-70 μm).

Schroederella delicatula (Peragallo)

Description: Pavillard Cells cylindrical and form chains; valves with depressions in the middle; valve ends with a crown of spines; a spine like pore canal present at center of each valve.

Dimension: Valve diameter, 14-41μm.

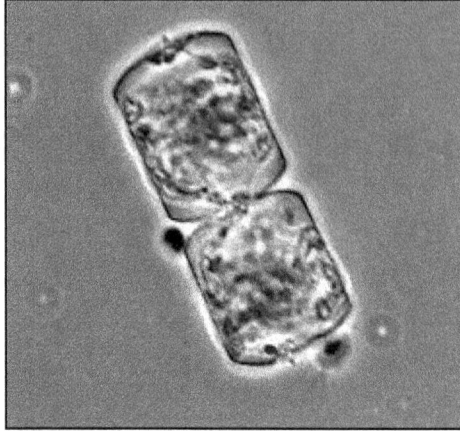

Odontella sinensis (Greville) Grunow

Synonym: Biddulphia sinensis.

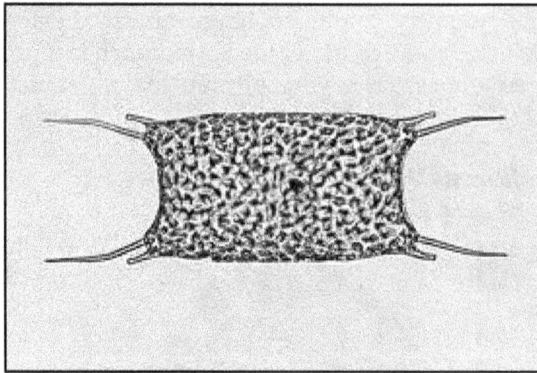

Description: Cells solitary or united by their spines to form loosely linked colonies. The frustule 2 times as long as it is broad. Weakly siliceous. Valves elliptical, having apices furnished with short processes. Valve surface slightly concave, bearing two spines which originate close to the base of the processes. Spines long, equal to or a little longer than the apical length of the valve. Chromatophores: numerous small irregular bodies scattered throughout the cell.

Dimensions: Apical axis of valve 120-260 µm; transapical axis of valve 60-80 µm; length of cell, up to 300 µm and 144-205 µm wide.

Biddulphia mobiliensis Bailey

Description: Cell resembles B. sinensis to some extent; cells moderately squarish with slender horns at corners of valves.

Dimension: Valve length of cell 24-81 µm.

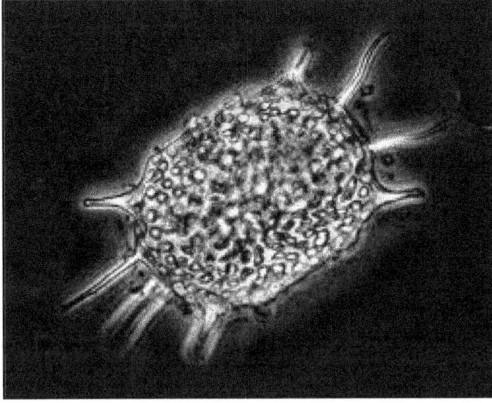

Triceratium reticulum Ehrenberg

Description: Concave margins. Middle part of the valve surface smoothly arched, apices separated by short furrows. Valve face areolated, areolae in the middle circularly-rectangular; irregularly distributed, towards the margin forming radial rows.

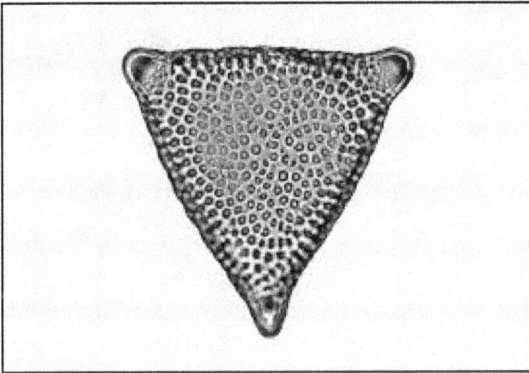

Dimensions: 25-80 µm long.

Cerataulina pelagica (Cleve) Hendey

Basionym: *Cerataulina bergonii.*

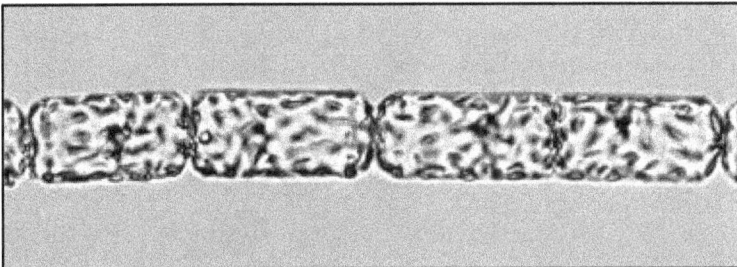

Description: Cells cylindrical, pervalvar axis usually twice or three times the diameter. Cells united to form chains, often twisted. Valves slightly convex, furnished with two short, stout cylindrical processes. Girdle composed of numerous intercalary bands, collar-like, Chromatophores: numerous rounded bodies, nucleus small, often pressed to one side against the cell wall.

Dimensions: Diameter of valve 36-56 μm; pervalvar axis 70-120 μm.

Climacodium frauenfeldianum Grunow

Description: Valve surface between elevations flat, apertures right angled to oblong and larger in pervalvar direction than in the cell proper Cells united into long, ribbon like chains. In girdle view with small, linear middle part with more or less long, thin processes on the poles of the apical axis.

Dimensions: Length of apical axis 70-90 μm, pervalvar axis 12-15 μm.

Eucampia zodiacus Ehrenberg

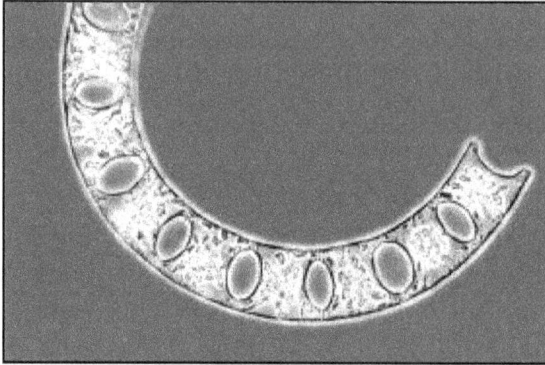

Description: Cells flattened, elliptical-linear in valve view, united in chains by two blunt processes. Chains spirally curved, with relatively narrow lanceolate or elliptical apertures. In girdle view, the intercellular apertures appear large, varying in shape from narrowly lanceolate to broadly elliptical. Chromatophores: numerous rounded bodies.

Dimensions: Length of polar axis 30-96 μm, pervalvar axis 40-50 μm.

Hemiaulus sinensis Greville

Description: Cells broadly elliptical in valve view. Chains straight or curved, more or less long. The Pervalvar axis more or less elongated. Valves with slightly convex surface of elliptical outline.

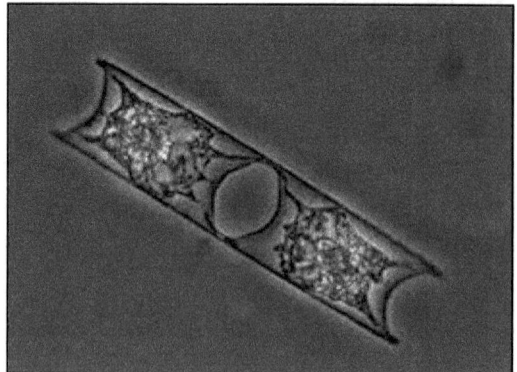

Dimensions: Apical axis 15-40 μm long.

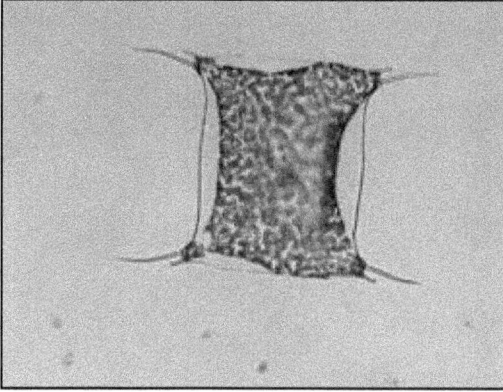

Bellerochea malleus (Brightwell) Van Heurck

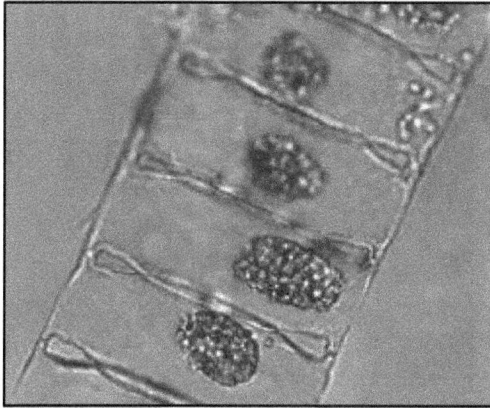

Basionym: *Triceratium malleus*

Description: Ribbons usually straight. Cells bi-angular, triangular, or quadrangular, united to form chains or flat colonies. The poles of the cell are raised to produce flattened processes when seen in girdle view in chain formation; the intercellular spaces are somewhat pear-shaped at each end of the cell. Valve surface slightly inflated in the middle. Chromatophores: numerous small plate-like bodies, scattered throughout the cell; nucleus central and usually distinct.

Dimensions: Length of valve 110 µm, pervalvar axis 20 µm; width 28.58-48.42µm.

Ditylum brightwelli (West) Grunow

Description: Cells triangular in space, somewhat like a prism, angles often rounded, giving a cylindrical appearance. Cells three to eight times longer than broad. Valves small, undulate and furnished with a corona of short but stout spines surrounding one large central spine. Central spine straight. Central area of valve often raised, hyaline. Girdle elongated. Chromatophores: numerous cocciform bodies, usually grouped towards the center of the cell.

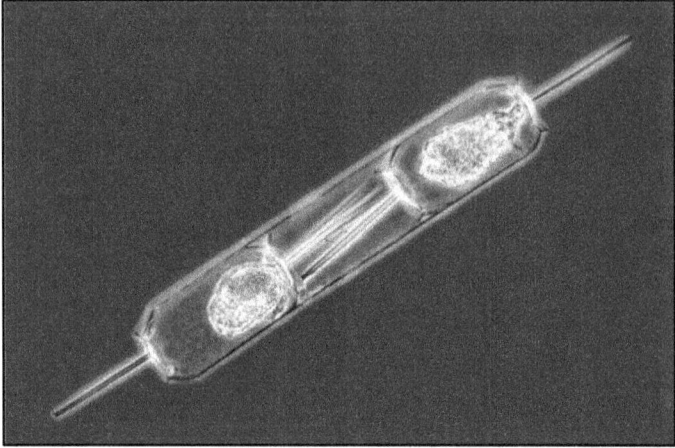

Dimensions: Diameter of valve 28- 46 µm; Pervalvar axis 80-130 µm; length of spine 20-50µm.

Corethron criophilum Castracane

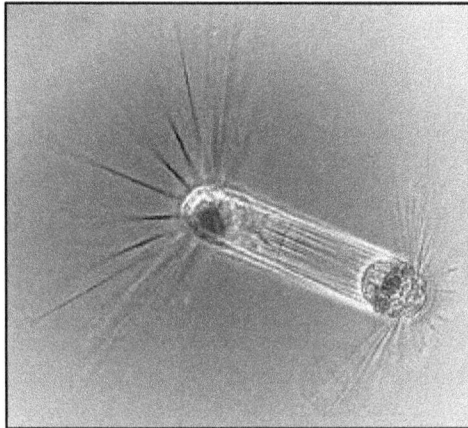

Synonym: *Corethron hystrix*

Description: Cells with cylindrical mantle and arched hemispherical valves. Circle of long slender setae at edge of valve. When free, on one valve all radiate out in the same direction from center of cell; on other valve two types of setae are found, longer ones of uniform width and approximately parallel to those of the first valve, and shorter ones ending in an irregularly twisted knob. Chromatophores numerous round or slightly elongated plates.

Dimension: Diameter 12-60µm, pervalvar axis 40-240 µm.

Rhizosolenia imbricata Brightwell

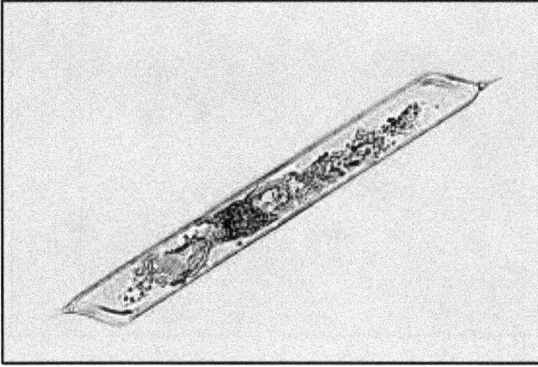

Description: Cells large, cylindrical, slightly flattened and furnished with depressed or flattened conical valves having a strongly oblique ventral margin. Valves furnished with a strong marginal spine, which appears as a continuation of the dorsal side of the valve. Girdle composed of two lateral rows of intercalary scale-like markings.

Dimensions: Diameter of valve up to 80 μm; length of cell up to 400 μm.

Rhizosolenia robusta Norman

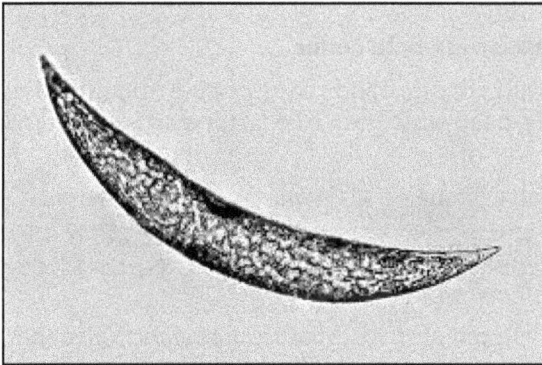

Description: Cells cylindrical, with deeply convex or conical curved valves, Valvar plane elliptical. Cells either crescent-shaped or s-shaped. Usually in single cellsngly, or in short chains. Intercalary bands robust, numerous, typically collar-shaped. Chromatophores: numerous, lying along the wall. Nucleus near the wall.

Dimensions: Diameter of valve 48-130 μm, length of cell up to 500 μm; length of conical valve up to 100 μm.

Rhizosolenia setigera Brightwell

Description: Cells cylindrical, tubular, straight and usually solitary. Valves deeply conical and produced to form an elongated spine. Spines usually straight, frequently occupying two-thirds of the total length of the cell. Girdle composed of two

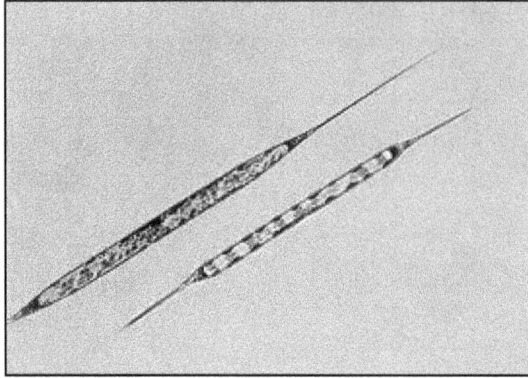

pervalvar lines of intercalary plates. Chromatophores: numerous small rounded bodies.

Dimensions: Diameter of valve up to 8-25 µm; length of cell up to 300µm.

Rhizosolenia cylindrus Cleve

Description: Cells cylindrical and valves with fairly truncated ends presence of large and bent spines; cell wall hyaline.

Dimension: Valve diameter, 21-24 µm.

Rhizosolenia crassispina Schroeder

Description: Cell cylindrical and valves possess truncated ends; apical processes broadened at the base and hair-like afterwards; numerous disc-shaped chromatophores.

Dimension: Valve diameter, 41-54 µm; length 145-278 µm.

Proboscia alata (Brightwell) Sundström

Basionym: *Rhizosolenia alata*

Description: Cells rod-shaped, cylindrical, straight; Valve shortly conical ending in tube-like. Intercalary bands scale-like, oeoo;wlwein two columns, numerous,

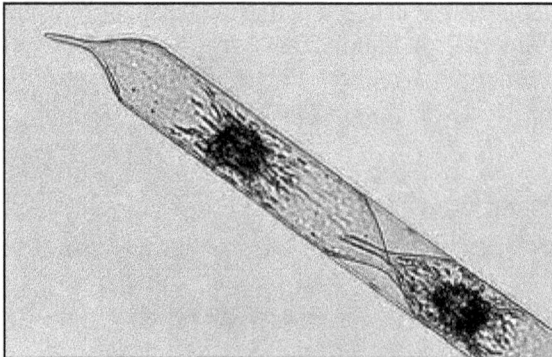

rhomboidal, with pores Proboscis, tip, truncate, short longitudinal slit just below tip. Cell wall thin, weakly siliceous Chromatophores: numerous, small.

Dimensions: Diameter of valve 7-18 µm, 1 mm in length.

Pseudosolenia calcar-avis (Schultze) **Sundström**

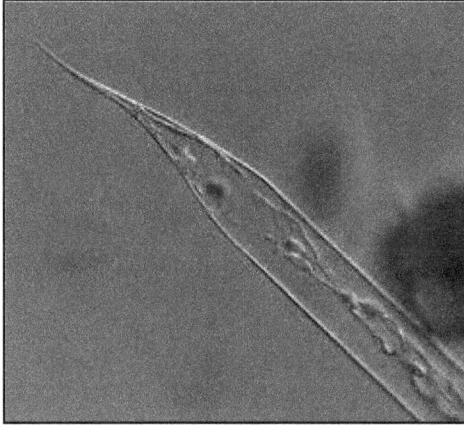

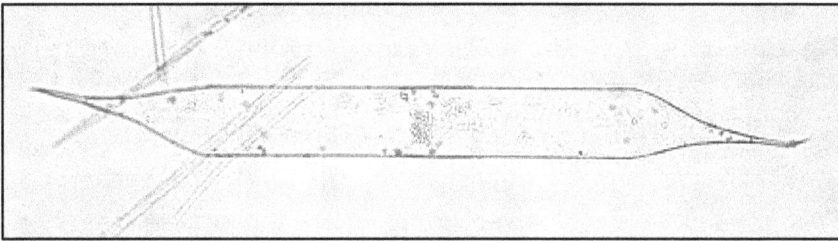

Basionym: *Rhizosolenia calcar-avis*

Description: Cells rod-shaped cylindrical, and the valves not so oblique, more regularly conical, curved at the apex. No wings on process. Cell wall thin and weakly siliceous very delicately punctuated. Chromatophores small, numerous.

Dimensions: Cells 6-53 µm in diameter, 1 mm in length, breadth 35-70 µm.

Guinardia flaccida (Castracane) **Peragallo**

Synonym: *Rhizosolenia castracanei*

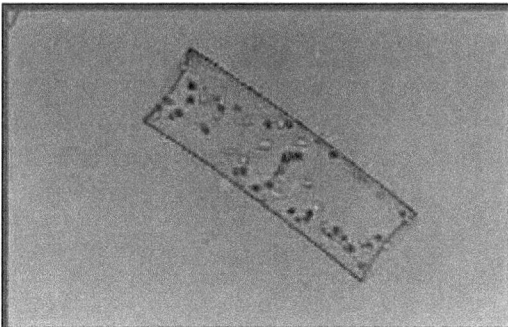

Description: Cells typically cylindrical, one and a half to several times longer than broad, single or united in chains by whole valve surface. Valve nearly flat, very slightly concave, with an irregular tooth at the margin Girdle composed of numerous intercalary bands; girdle hyaline, weakly siliceous. Chromatophores: lying near the wall, round shaped, more or less lobed nucleus more or less central.

Dimensions: Diameter of valve 36-80 µm; pervalvar axis of cell, 160 µm.

Bacteriastrum delicatulum Cleve

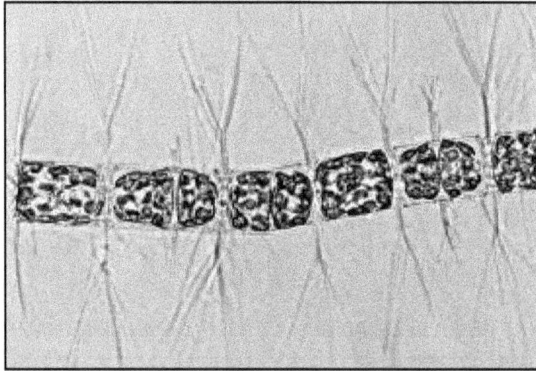

Description: Cells cylindrical. Chains long, straight. Setae 6-12, with strong, long basal part. Forked parts slightly curved, smooth or somewhat wavy. Terminal setae of both ends directed toward the inside of the chain and in front view of the valve similarly curved; stronger than inner setae and with fine spines arranged spirally. Chromatophores: small, numerous, distributed along cell wall.

Dimensions: Diameter of valve 6-15 µm), 12-20 µm, pervalvar length 20-60µm. Chains of up to 20 cells.

Chaetoceros affinis Lauder

Description: Cells united to form short, straight chains, cylindrical, rectangular in girdle view. Valves elliptical, valve surface flat or weakly convex, with slightly produced apices leading to the thin and almost straight setae that emerge from the

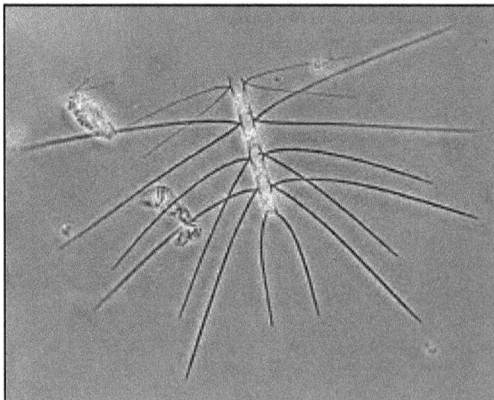

valve margin almost at right angles. Chromatophore, a single body lying against the girdle with a large central pyrenoid and occupying the major part of the cell. Nucleus central. Upper and lower surfaces almost equal and bearing numerous spines.

Dimensions: Chains, 7-27 µm wide, diameter of valve 10-25 µm.

Chaetoceros curvisetum Cleve

Description: Cells united to form spirally twisted chains, without distinct terminal cells, rectangular in girdle view. Valves elliptical, with concave surfaces. Cells four-cornered in broad girdle view, adjacent cells connected by conspicuous corners. Setae arising from corners of cells, all bent towards the same side of the chain-toward outside of the curved axis of spiral. Setae long and slender. Chromatophores one per cell with large central pyrenoid.

Dimensions: Chains, 7-30 µm wide); apical axis 10-30 µm.

Chaetoceros lorenzianus Grunow

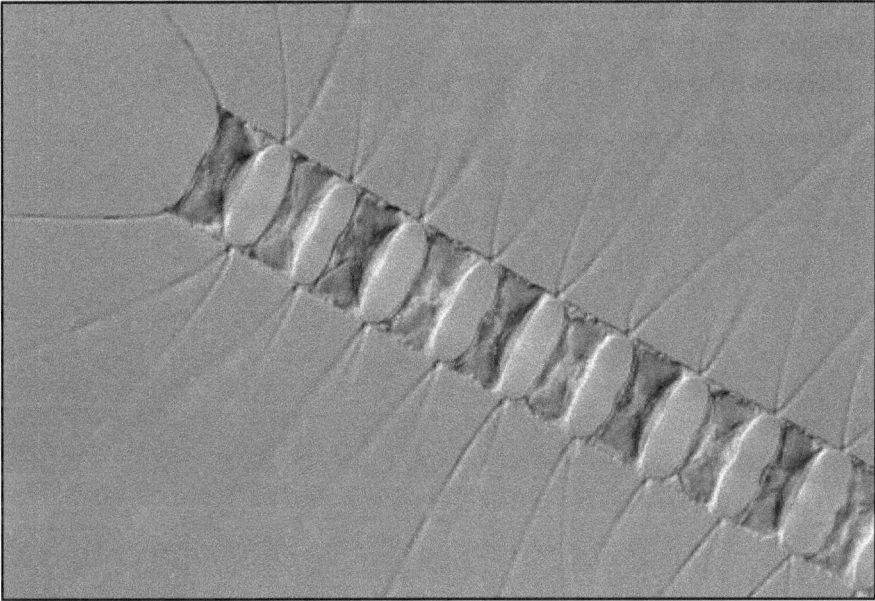

Description: Cells generally united to form short chains, but often solitary, rectangular in girdle view. Valves elliptical, central area flat or slightly convex. Valve mantle usually deep, with a narrow girdle. Apices of the valves produced to form long stiff setae that emerge immediately from the valve margin and fuse with those of the adjacent cell at the point of exit only. All setae are divergent and lie more or less in the same plane, but setae of upper and lower terminal cells are strong, divergent and thicker or wider in the more remote half of their length. Chromatophores: large, platelike, 4-10 plates.

Dimensions: Cells 7-48 µm wide diameter of valve 26-60 µm.

Chaetoceros didymus Ehrenberg

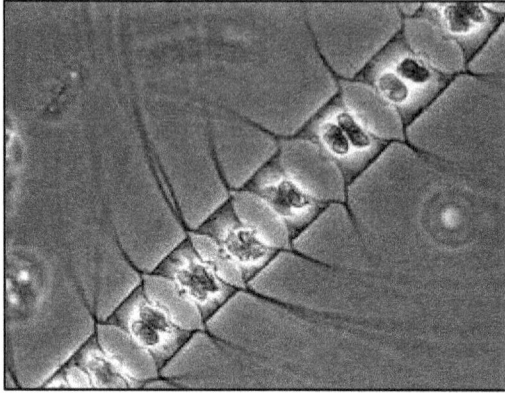

Description: Cells from straight chains; a characteristic semi-circular knob like structure present in middle of each valve; distinct interlocking of setae, two plate-like chromatophores present.

Dimension: Length of the cell, 22-40 µm.

Chaetoceros diversus Cleve

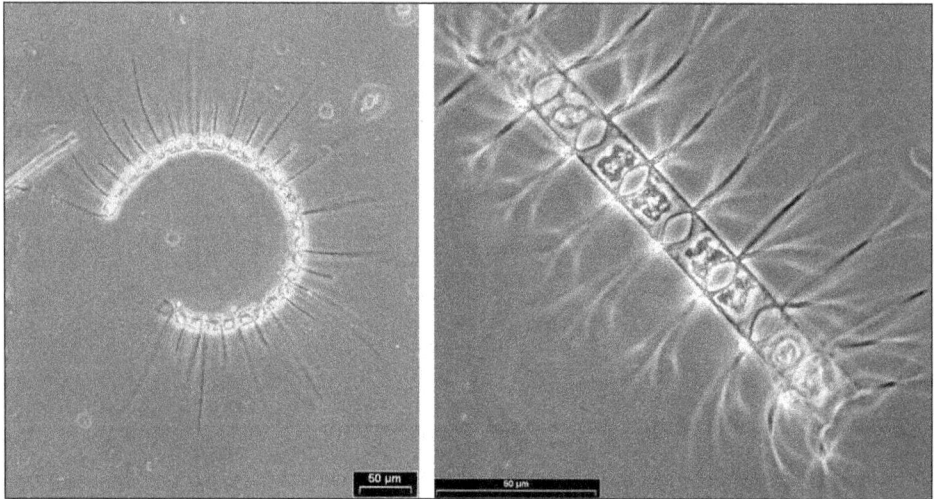

Description: Cells form compact and short chains; apertures very small; setae of some cells thicker, tubular and spinous; other inner setae and terminal ones hair-like.

Dimension: Length of cell, 5-9µm.

Leptocylindrus danicus Cleve

Description: Cells, tubular, cylindrical, narrow, two to ten times as long. United in closed, long, straight, stiff chains. Valves circular, flat or convex, occasionally concave, without visible sculpturing. Girdle elongated and composed of numerous

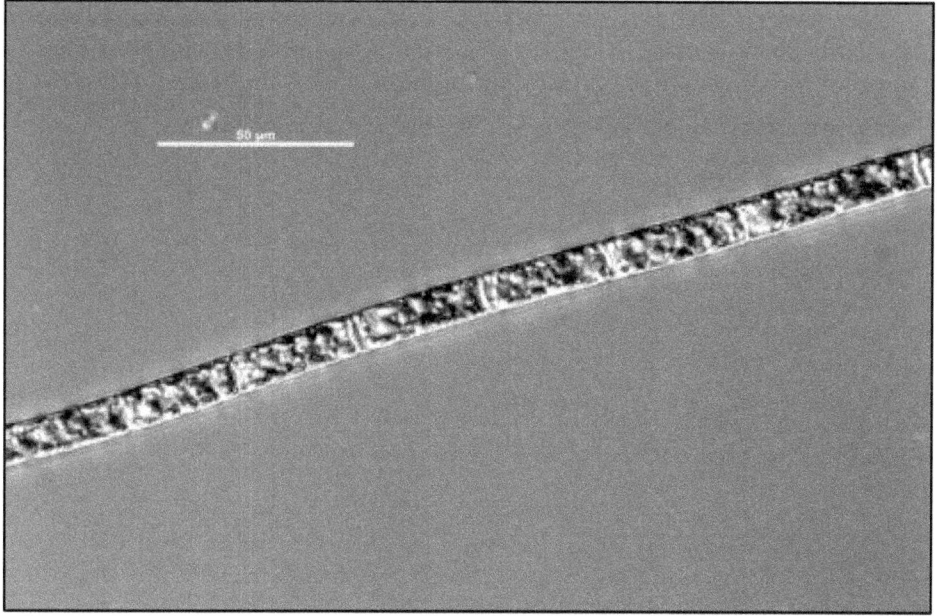

pointed intercalary segments. Adjacent cells often with only one cell wall between two valves. Chromatophores: few to numerous, not very small, oval plates, distributed throughout the cell.

   Dimensions: Cells 7-10 µm, 5-16 µm in diameter, pervalvar axis 30-50µm.

Leptocylindrus minimus Gran

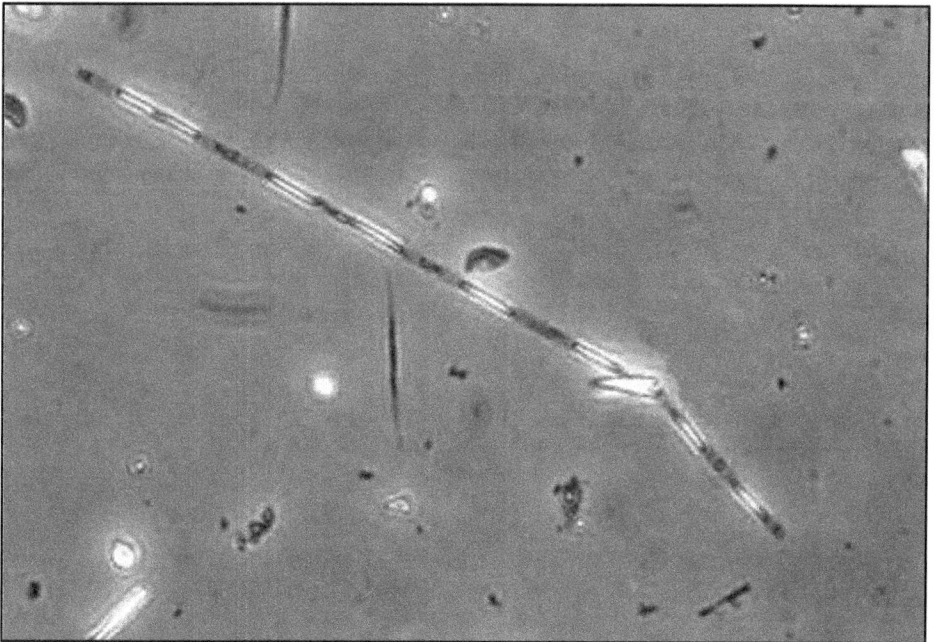

Description: Cells tubular chlorophyll consists of only two (seldom one) elongate plate-like bodies, often lying along the girdle. Girdle composed of numerous narrow intercalary bands having pointed or cuneate ends.

Dimensions: Diameter of valve 5-6 µm; pervalvar axis 40-50 µm.

Order: Pennales

Asterionellopsis glacialis (Castracane) Round

Basionym: Asterionella glacialis

Synonym: Asterionella japonica

Description: Cells united to form spiral star-shaped colonies, eight to twenty cells to a colony. Cells having one end inflated into a triangular head, while the other end is produced into a narrow rod-like outer portion. Valves shaped somewhat like a cricket bat, with one broad flattened end and a linear handle-like extension. Valve possessing a narrow pseudoraphe. Chromatophores: usually two, confined to the broad end of the cell.

Dimensions: Length of valve 30-150 µm; length of enlarged region 10-23 µm; width of enlarged part 8-12 µm.

Opephora schwartzii (Grunow) Petit

Synonym: Fragilaria schwartzii

Description: Frustules small, rectangular in girdle view. Valves linear, elongate, and slightly clavate, with a rounded upper apex, slightly convex margins and a smaller, narrower but rounded lower apex. Valve surface furnished with coarse areolae at right angles to the narrow pseudoraphe placed in the apical axis.

Dimensions: Length of valve 60-100 µm, breadth 10-12 µm.

Thalassionema nitzschioides (Grunow) Mereschkowsky

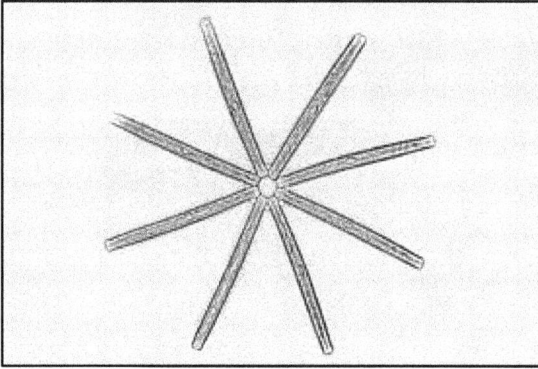

Synonym: Thalassiothrix nitzschioides

Description: Cells united to form stellae or zigzag colonies. Cells in girdle view narrow linear, often slightly curved. Valves narrow linear with parallel sides and blunt-rounded ends. Valve surface structureless, but the margins are furnished with minute puncta. Chromatophores: numerous cocciform bodies.

Dimensions: Length 30-80 µm; width 2-3.5 µm.

Thalassionema frauenfeldii (Grunow) Hallegraeff

Synonym: Thalassiothrix frauenfeldii

Description: Cells united into star-shaped colonies or zigzag bands. In girdle view linear. Valves very narrow linear, ends distinct but only slightly unlike, one end

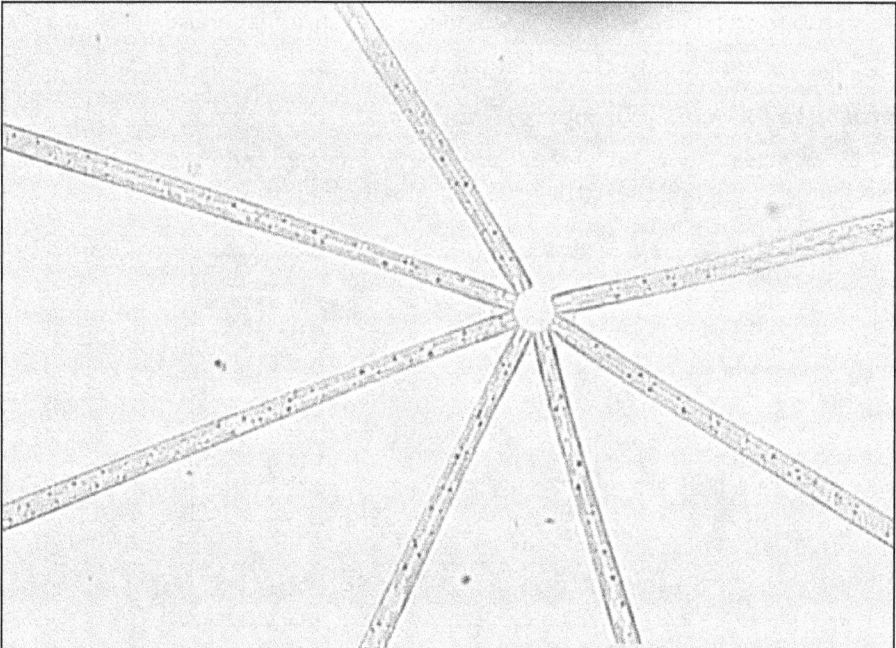

blunt-rounded, near the other end usually widened then decreased to form a wedge-shaped point. Valves structure less.

Dimensions: Valves 90-120 µm long, 2-4 µm wide.

Thalassiothrix longissima Cleve and Grunow

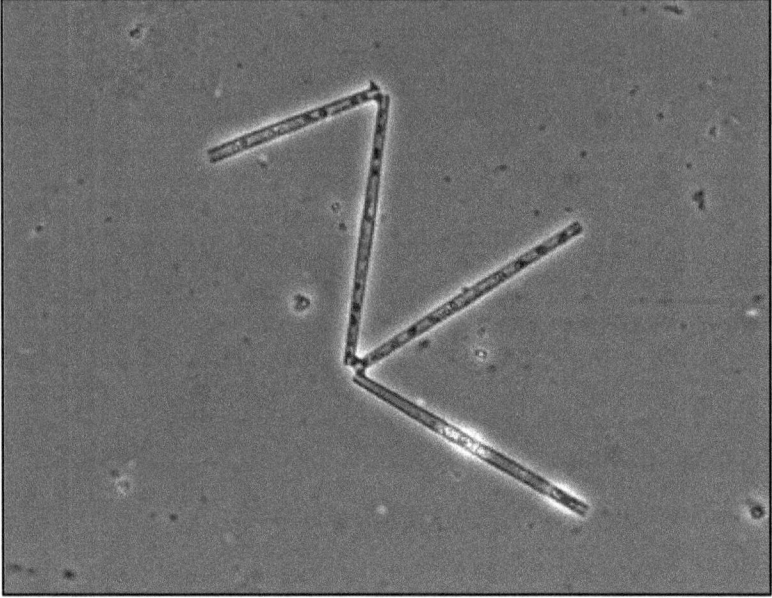

Description: Cells usually free, but frequently matted together in dense masses. Cells very long, thread-like, usually slightly bent, rectangular, or prismatic in transapical section. Valves elongated, linear, with dissimilar ends.

Dimensions: Width 3-6 µm, length of valve 1-4 mm.

Gyrosigma balticum (Ehrenberg) Rabenhorst

Description: Valves linear with curved and truncated ends; raphe excentric and central area small, oblique; transverse and longitudinal striae equidistant, 11-12 in 10 µm.

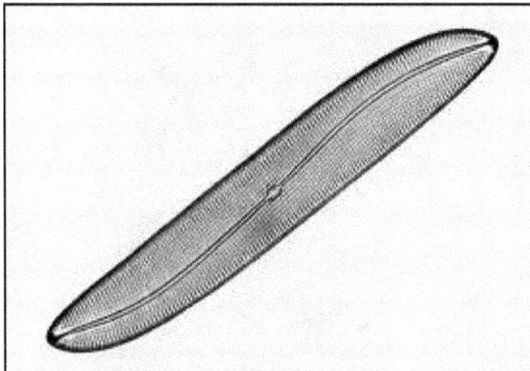

Dimension: Valve length of cell 290-338 µm; breadth 28-30 µm.

Pleurosigma galapagense Cleve

Description:Valves very slightly sigmoid; ends blunt; raphe somewhat sigmoid; transverse striae 18 in 10 µm and oblique striae 15 in 10 µm.

Dimensions: Width = 1.2-5.5um, Length = 64-160um.

Diploneis weissflogii (A. Schidt) Cleve

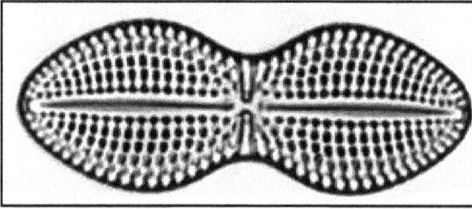

Description: Valves broad and strongly constricted at center; ends fairly rounded; central nodule with horns; transverse costae 9 in 10 µm.

Dimensions: Valve length 28-58 µm; breadth 10-25 µm (away from constriction) and 6-15 µm (at constriction).

Navicula directa Cleve

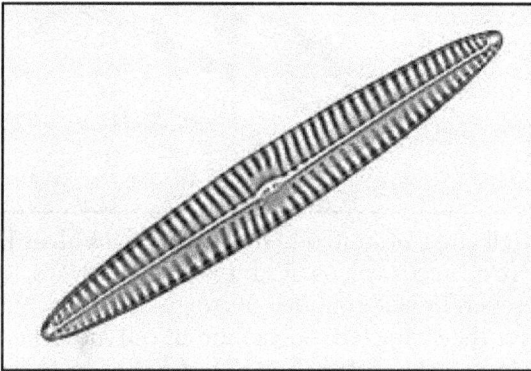

Description: Valves long, slender, rhombic-lanceolate, with sub-acute apices. Axial area narrow, polar and central nodules small. Central area moderately large, sub-rectangular, equal to half the valve width. Valve surface striate, striae well-shaped, coarsely lineate, very weakly radiate at the center, transverse throughout the remainder of the valve.

Dimensions: Length of valve 150-180 µm; breadth 15-17 µm.

Pleurosigma elongatum Smith

Description: Valves elongate, linear-lanceolate, gently tapering from about the middle to acute apices. Valves gently sigmoid, raphe central, central and polar nodules small.

Dimensions: Length of valve 130-380 µm, breadth 24-30 µm.

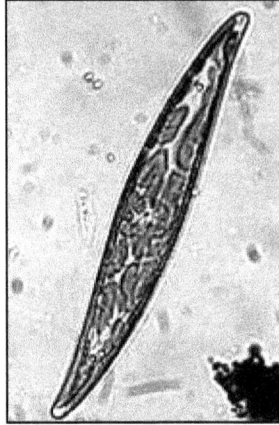

Bacillaria paxillifera (O. F. Müller) Hendey

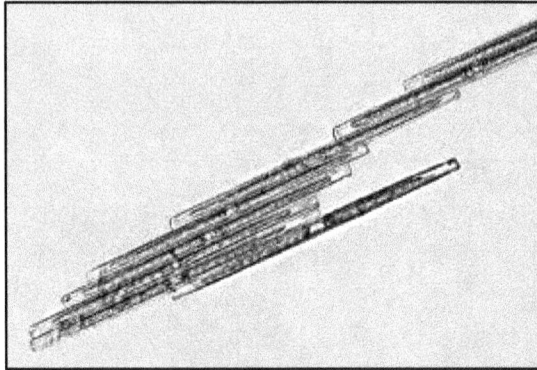

Description: Cells free, or united to form a colony. Valves linear-lanceolate to lanceolate, with produced apices. Valve surface striate; striae transverse. Chromatophores: several small, rounded or irregular bodies. Nucleus central. This species is known for the characteristic method of cell movement. The cells at one moment appear to form a filament or short ribbon with adnate frustules, then with a gliding movement the cells slide forward, adhering one to another only by their ends, to form an elongated chain.

Dimensions: Length of valves 80-115 µm; width 5-6 µm.

Nitzschia longissima (Brébisson) Ralfs in Pritchard

Description: Valves linear-lanceolate tapering to very long apical extremities. Valves usually straight over the whole length, extremities not curved. Raphe with fibulae.

Dimensions: Length of valve 125-250µm; apical axis, 125-450µm; transapical axis 300-400µm.

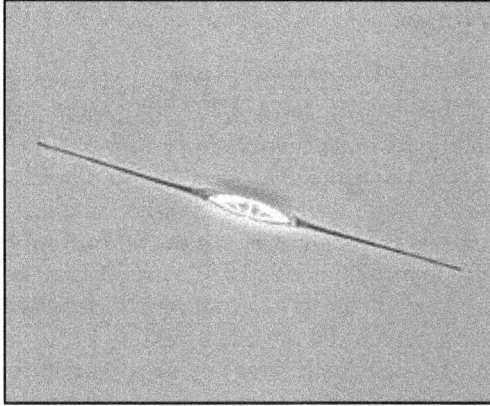

Nitzschia sigma (Kutzing) W. Smith

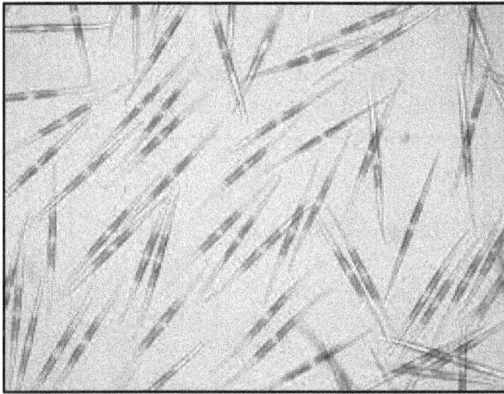

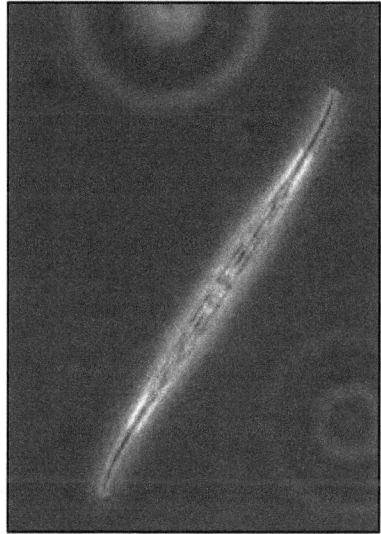

Description: Valve linear; somewhat sigmoid in girdle view and straight in valve view; bulge at the center and gradually diminishing in size towards end; keel punctae 5-6 in 10µm.

Dimensions: Valve length, 280-310 µm; 10-11 µm.

Nitzschia closterium (Ehrenberg) W. Smith

Description: Valve spindle shaped in middle; extremities beak like and slightly curved in opposite directions; striae not visible.

Dimensions: Valve length 30-160 µm; breadth 3-7 µm.

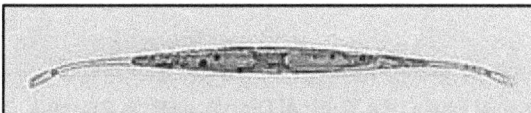

10.2. Dinoflagellates

Class: Pyrrophyceae (Dinoflagellates)

The dinoflagellate species are generally small (less than 100 µm), unicellular flagellates inhabiting marine or brackish water environments. They may be planktonic, benthic or epiphytic, and some are obligate parasites. Dinoflagellates are characterized by the presence of two dissimilar flagella, one is ribbon shaped encircling the cell, the transverse flagellum, and the second is whip-like trailing behind the cell, the longitudinal flagellum. In most species, flagella are situated in furrows on the cell body, namely the transverse furrow (cingulum), and the longitudinal furrow (sulcus). The side of the cell from which the flagella originate is the ventral side. In a few species of prorocentroids, the flagella arise anteriorly and the cingulum and sulcus are absent. The transverse flagellum helps in providing propulsion and the longitudinal flagellum is for direction. Most dinoflagellates have a distinct nucleus. Dinoflagellates may be divided into armoured (thecate) and unarmoured (non-thecate or naked) species based on the presence or absence of thecal plates. The armour of dinoflagellates is divided into an upper (epicone or epitheca) and a lower (hypocone or hypotheca) half, and consists of polygonal plates, which fit tightly against each other. The important constituent of the armour is a polysaccharide (cellulose). The taxonomy of armoured dinoflagellates is primarily based on the arrangement and number of the thecal plates, whereas the taxonomy of the unarmoured species is mainly based on cell shape and cingular and sulcal arrangement.

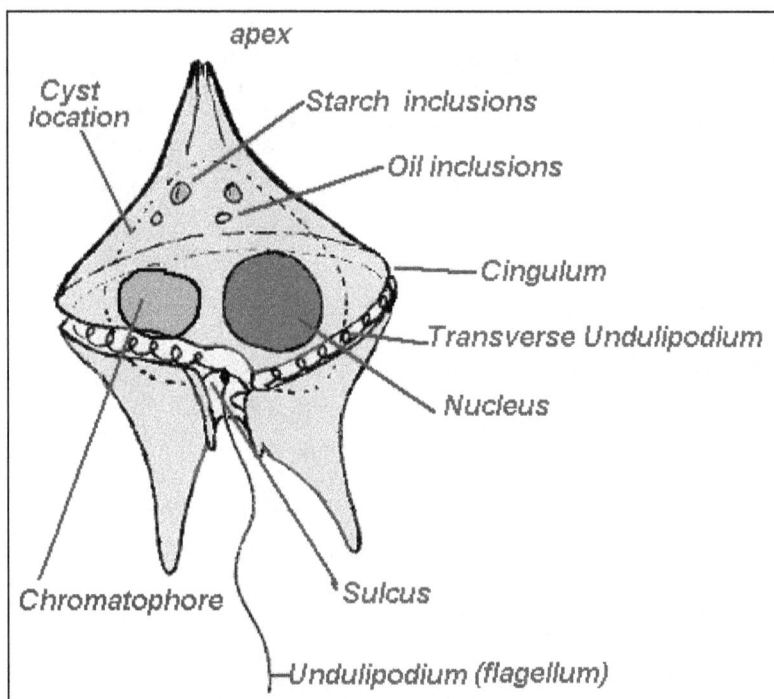

Internal Parts of a Typical Dinoflagellate Species.

Group: Desmophyceae

Order: Prorocentrales

Prorocentrum micans Ehrenberg 1833

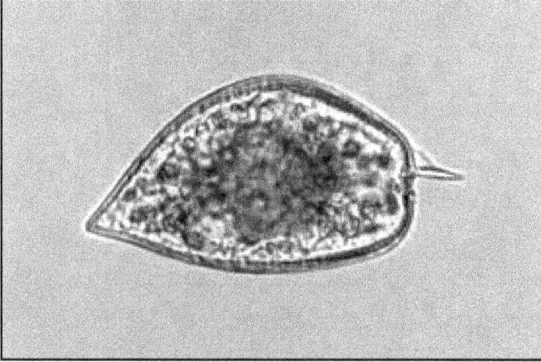

Description: Medium-sized, pyriform to heart-shaped cell with a rounded anterior end and pointed posterior end, broadest around the middle, usually less than twice as long as broad. Prominent anterior spine with wings. Thecal plates mainly arranged in radial rows, and surface of plates covered by regularly arranged depressions. Chloroplasts containing large internal pyrenoids. Nucleus large, V-shaped and situated in the posterior end of the cell.

Dimensions: Length 35-50 μm.

Prorocentrum rostratum Stein

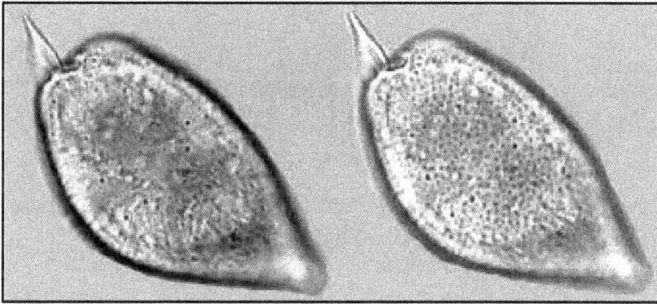

Description: Body compressed laterally with a blunt apex; a finger or rostratum like process present; valves narrow and pointed posteriorly with an apical tooth on each valve.

Dimension: Valve length, 98 μm; breadth 18μm.

Exuviaella compressa Barley and Ostenfeld

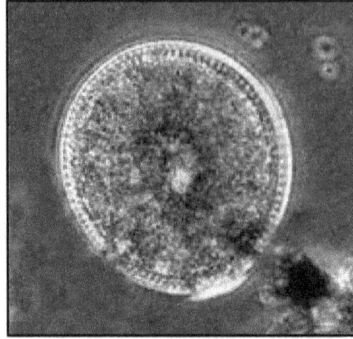

Description: Cell oval and not compressed; each valve with a smooth tooth anteriorly; two plate-like yellow chromatophores.

Dimension: Valve length 20-25 µm; breadth 18-24 µm.

Phalacroma argus Stein

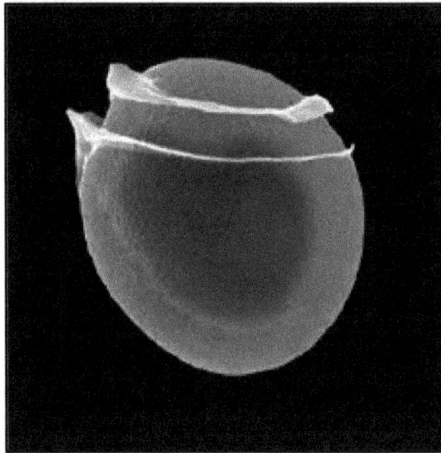

Description: Body laterally ovate and wider behind girdle; epitheca and hypotheca rounded; girdle lists ribbed; right sulcal list concave.

Dimensions: Length 72 µm.

Phalacroma cuneus Schutt

Description: Body cuneate laterally; epitheca low and broadly rounded; hypotheca posteriorly narrowly rounded to subacute; margin of left sulcal list slightly sigmoid 71 µm.

Dimensions: Length, 70-170 µm.

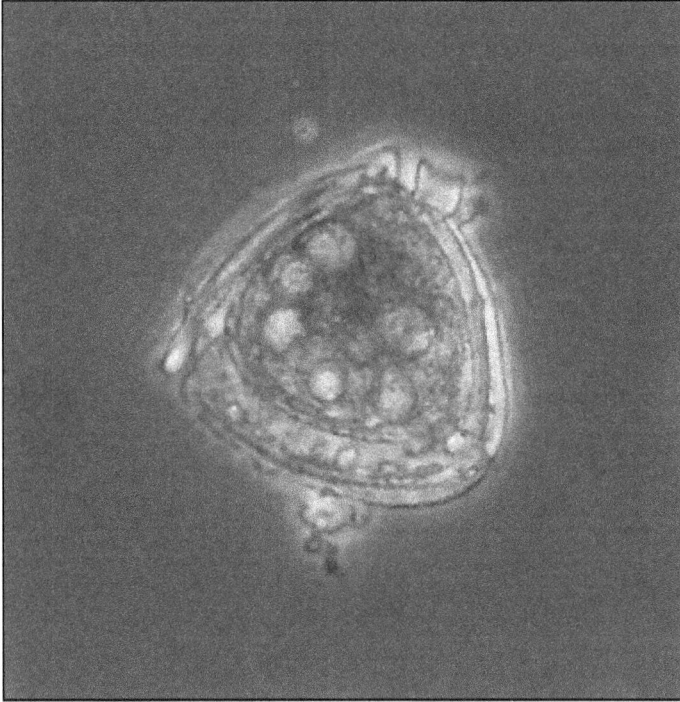

Group: Dinophyceae

Order: Dinophysiales

Dinophysis caudata Saville-Kent 1881

Description: Medium-sized species with a characteristic posterior finger-like process; cells often occur in pairs, dorsally attached. Body very variable; epitheca low; hypotheca long, widest at or near the middle. The epitheca is obscured by a deep funnel formed by the anterior cingular list and supported by ribs. At the distal margin the funnel flattens out. The posterior cingular list is narrow dorsally, becoming wider

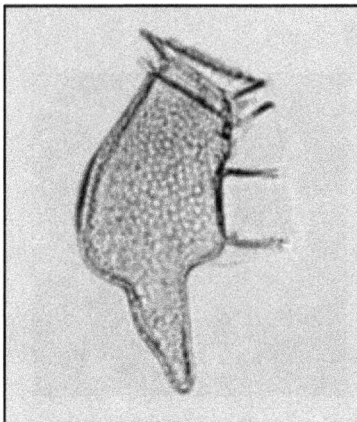

on the left ventral side where it meets the left sulcal list. The posterior cingular list is projected anteriorly and is supported by ribs. Viewed laterally, the hypotheca is narrow at the girdle; the ventral margin is usually straight or sigmoid to the base of the left sulcal list. The hypotheca has its widest point usually at the position of the base of the left sulcal list. The dorsal side of the hypotheca may curve sharply towards the centre. The epithecal and hypothecal plates are covered with areoles each containing a pore or, at the plate margins, one or two pores. The left sulcal list is wide and supported by three ribs spaced equally apart, which are straight or slightly curved. The lists may be reticulated. Twinned forms occur frequently, joined together on the dorsal surface at the widest part of the hypotheca.

Dimensions: Length, 70-170 μm.

Dinophysis miles Cleve

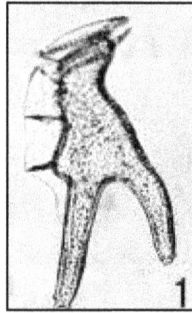

Description: This species is known for its extreme body modifications.

Dimensions: 150 μm long, 60 μm wide.

Ornithocercus magnificus Stein 1883

Description: Small to medium-sized circular full body in lateral view with extensive sulcal and cingular list and rib systems which characterize the species. Ribs and lists are formed at the extremities of plates, near sutures. Body surface markings of pores, poroids, or areolae. Chloroplasts are absent but photosynthetic symbionts may be present in cingular chamber.

Dimension: Length 40-120 μm.

Order: Gymnodiniales

Akashiwo sanguinea (Hirasaka) Hansen and Moestrup 2000

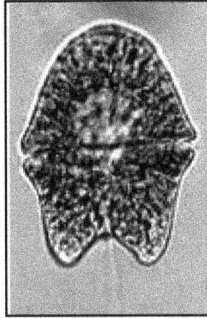

Basionym: *Gymnodinium sanguineum.*

Synonyms: *Gymnodinium splendens*

Description: Cell ovoid, flattened dorso-ventrally with convex dorsal surface and concave ventrally. Epicone and hypocone nearly equal, in ventral view the epicone is in the shape of a blunt, rounded cone and the hypocone is deeply indented by the sulcus. The girdle is impressed, left-handed displaced one to two girdle widths. The sulcus expands posteriorly. The nucleus is slightly anterior of the midline. Numerous yellow-brown chromatophores radiate from outside the centre to the periphery. Longitudinal flagellum is about one and a quarter times the body length.

Dimension: 40-80 µm long.

Karenia brevis (Davis) Hansen and Moestrup 2000

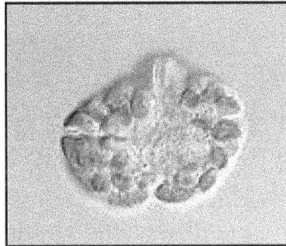

Synonym: *Gymnodinium breve*

Description: Dorsal and ventral views of this species show bulbous apical carina. Ventral view with, peripheral chloroplasts, the nucleus in left hypotheca and the open ended sulcus onto epitheca, adjacent to the straight apical grove.

Dimensions: 19 µm wide.

Order: Peridiniales

Ceratium furca (Ehrenberg 1836) Claparéde and Lachmann 1859

Description: Body straight; epitheca tapering gradually into apical horn;

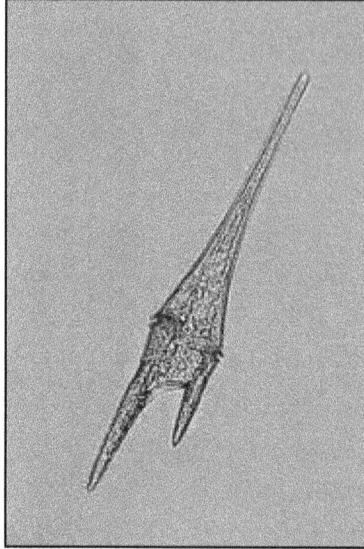

hypotheca varying from parallel-sided to tapering from girdle; antapicals strong, unequal, usually straight, parallel to subparallel, may be toothed.

Dimension: Length 70-100 μm

Ceratium fusus (Ehrenberg 1834) Dujardin 1841

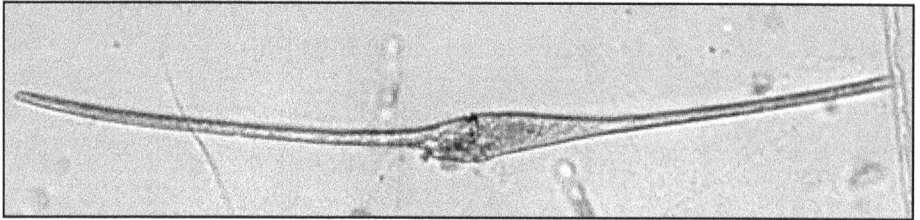

Synonym: *Peridinium fusus*

Description: An elongated needle-shaped species. Medium to small species, widest point adjacent to the girdle from which the epitheca long, gradually tapering to a cylindrical or gently tapering apical horn, usually slightly bent dorsally or straight; hypotheca tapering; left antapical long, slightly curved, rarely straight; right antapical rudimentary or absent.

Dimensions: Length 200-300 μm, 15-30 μm width.

Ceratium trichoceros (Ehrenberg) Kofoid 1908

Description: Body small; epitheca rounded; horns thin and long; antapicals beginning almost parallel to girdle, then curved until they are parallel with apical horn and almost the same length forming a flat-bottomed U.

Dimension: Length 300-500 μm.

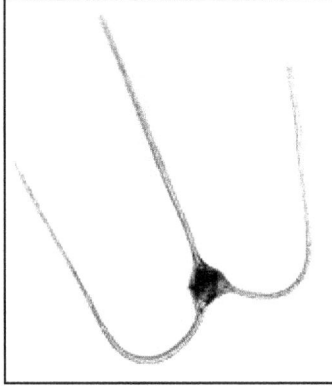

***Ceratium tripos* (O. F. Müller 1781) Nitzsch 1817**

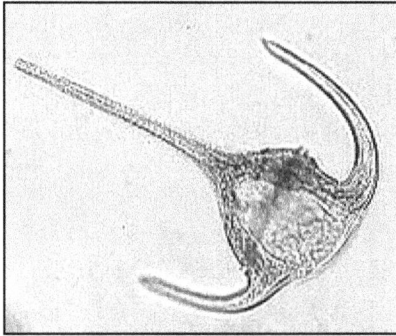

Description: Body of cell stout, about as broad as it is long. Epitheca triangular leading sharply into a fairly long straight horn, plates usually clear. Antapicals more or less continuous with slightly flattened base, then sharply curved forward, parallel to or making an acute angle with apical horn. Hypotheca slightly flattened at the posterior end.

Dimension: Length 75-90 µm

***Ceratium pulchellum* B. Schroder**

Description: Robust species; eiptheca with steep left and very convex right side; apical horn long and strong, slightly wider in middle; base of hypotheca strongly convex; antapicals short; less stronger than apical; left curved slightly diverging or parallel to apical; right equal in length or shorter than left.

Dimensions: Length of horns- left, 130 – 140um; right, 60-65 µm.

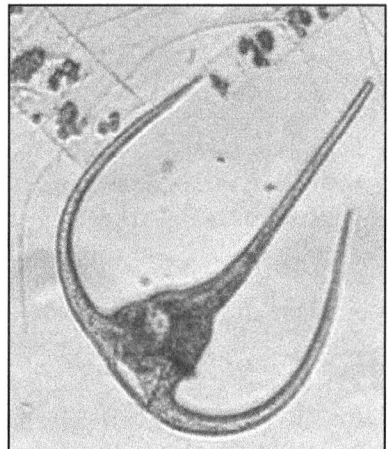

Ceratium breve (Ostenfeld and Schmidt) Schroder

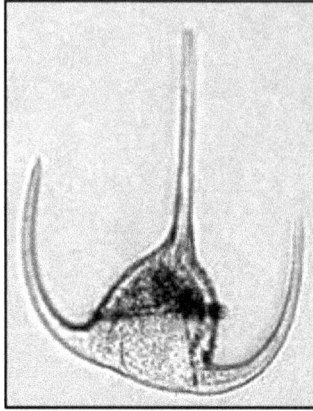

Description: Body with short horns; epitheca's right contour strongly convex and left contour steep; hypotheca with evenly convex base; anatapicals very strong; slightly parallel with apical horn.

Dimensions: Length of horns- antapicals, 60um; apical, 75-80μm.

Ceratium karstenii Pavillard

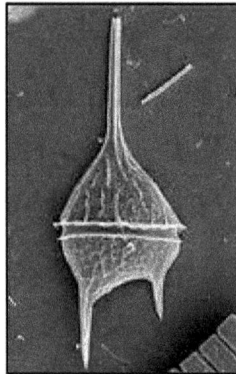

Description: Strong body; right contour of the epitheca convex; apical horn slightly bent at base; antapical slender; right antapical longer than left antapical and bent distally towards apical horn; left anatapical at times curved.

Dimensions: Length of horns- antapicals, 25-32 um;apical, 90-95μm.

Ceratium contortum (Gourret) Cleve

Description: Cell resembles C. karstenii to some extent; epitheca oblique on right, right contour strongly convex; apical horn twisted, S shaped; horns slender; antapicals unequal, right longer than left and twisted towards apical horn.

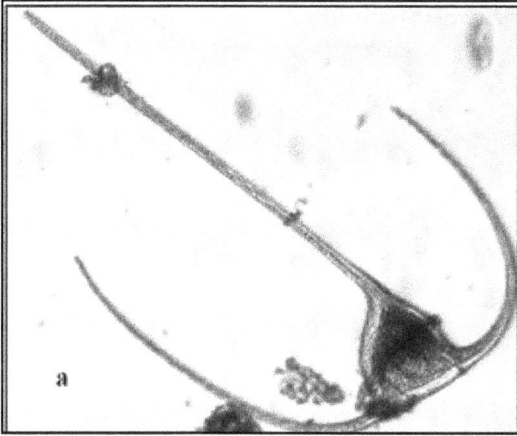

Dimensions: Length of horns- antapicals, 25-30 um;apical, 90-94µm.

Protoperidinium depressum (Bailey 1855) Balech

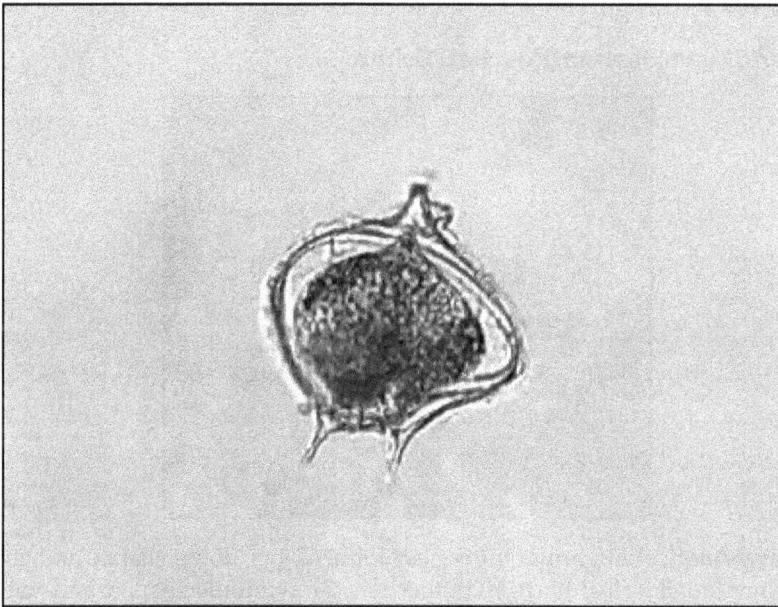

Synonym: *Peridinium depressum*

Description: Cell broad, obliquely flattened dorsoventrally; axis very oblique. Epitheca tapering into large apical horn, whereas hypotheca with two long, tapering antapicals, each with a spinelet continuous with the sulcal list. The sulcus is deeply excavated. Girdle left-handed, unexcavated, bordered by wide lists supported by spines. Cell contents pink. A bioluminescent species.

Dimensions: Length 100-200 µm; 116-144 µm wide.

Protoperidinium murrayi Kofoid

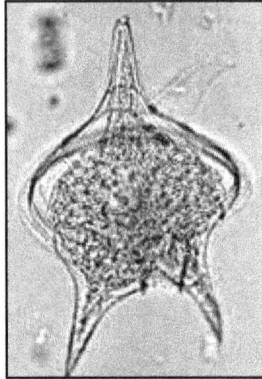

Synonym: Peridinium oceanicum

Description: The thecal plates are smooth; apical horn much elongated and arises more abruptly from the epitheca. The antapical horns are more divergent. The girdle is clearly at the widest point of the cell.

Dimensions: Length about 200 µm, width 100 µm.

Protoperidinium ovatum (Pouchet) Schutt

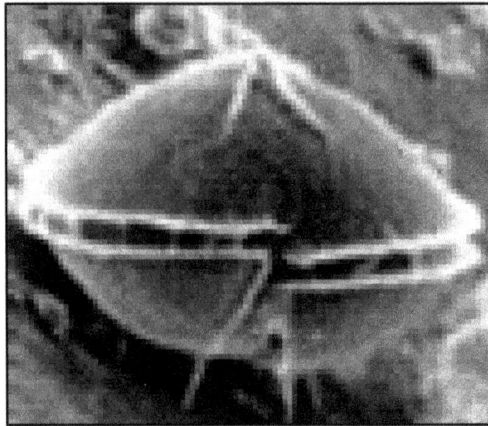

Description: Cell slightly compressed; epitheca low, dome-shaped and tapering sharply into small apical horn; hypotheca also low, dome-shaped and with two small antapical spines; sulcus subantapical or reaching antapex; five-sided 1st apical and four-sided or five-sided 2nd intercalary.

Dimensions: Length 68-70µm, width 45-60 µm.

Podolampas palmipes Stein 1883

Description: Cell pyriform, elongate, pear-shaped, narrow in front. The epitheca is drawn out into a long, slender neck. The intercalary plate is more or less pentagonal. The precingular plates bear elongated pores roughly arranged in a transverse row.

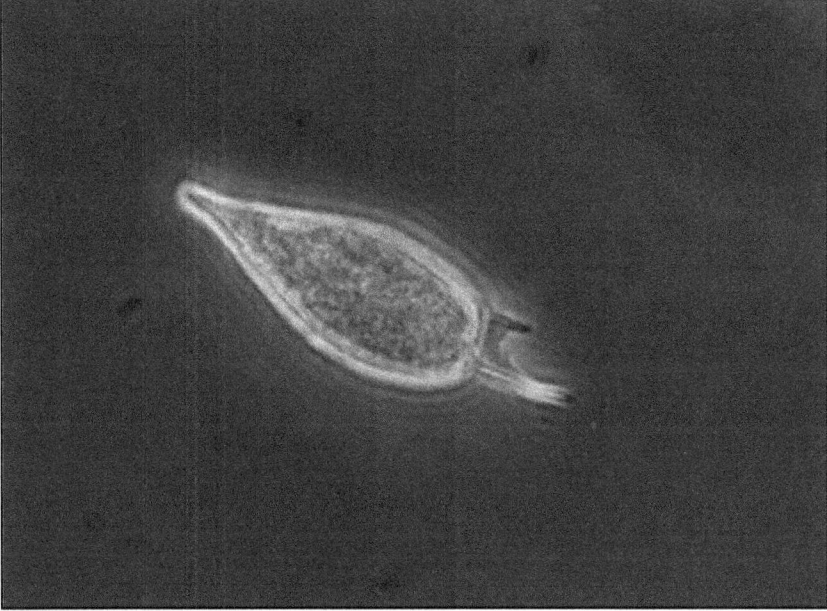

There are three cingular plates. The hypotheca is greatly reduced in depth being less than one tenth of body length. The right hand spine is much shorter than the left antapical spine. The nucleus is elongated and situated in the right half of the cell in the cingulum region.

Dimensions: Length 70-110 µm, 20-37 µm wide. Length of left spine 4-28µm; length of right spine 14-23 µm.

Order: Gonyaulacales

Pyrophacus horologicum Stein 1883

Description: Cell low, biconvex; lens-shaped test with epitheca and hypotheca equal, much wider than high; plates well marked with linear markings in the middle; sulcus with a few small plates. Cytoplasm with a strong tendency to round up, containing numerous chloroplasts.

Dimension: Length 40 µm; width 35-136 µm, height 32-125 µm.

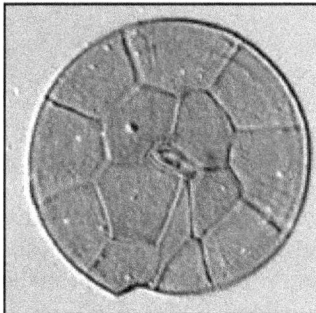

10.3. Other Groups

Class: Cryptophyceae

Plagioselmis sp.

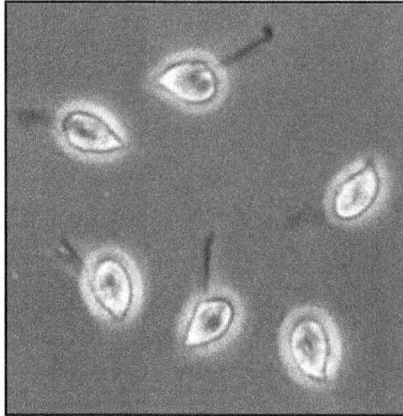

Cell, small, asymmetrical with furrow or depression. Gullet or furrow lined with two or more rows of ejectosomes. Two flagella, originating at the end of the furrow / gullet, with two and one row of fine tubular hairs Mode of swimming is heterodynamic. Its color brown, green, red, or blue. One or two chloroplasts.

Dimensions: 6-10 μm.

Class: Raphidophyceae

Heterosigma sp.

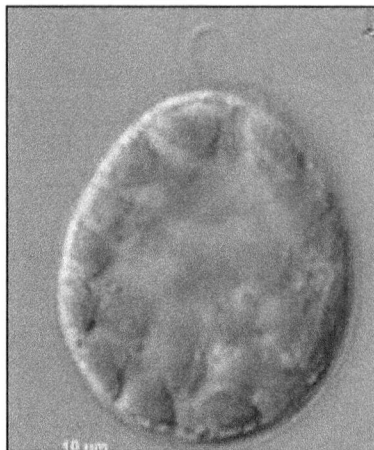

Cells are symmetric, more or less flattened. A more or less pronounced flagellar groove may be present. Cell surface smooth, with two flagella, one flimmer flagellum, pointing forward, the other also with hairs, often in shallow ventral groove, pointing backward. Colour yellow to yellowish brown (due to fucoxanthin), many discoid chloroplasts.

Class: Dictyochophyceae

Dictyocha speculum Ehrenberg 1839

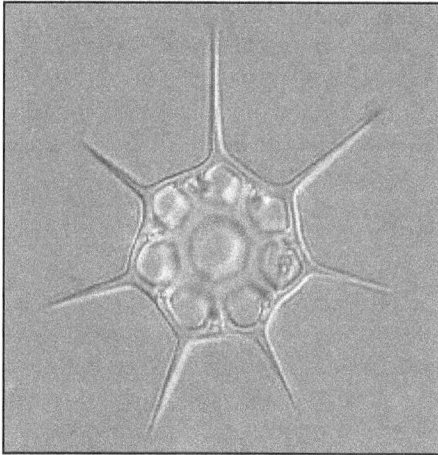

Synonym: *Distephanus speculum*

Description: It has hexagonal skeleton, with many yellow-brown chloroplasts.

Dimensions: Skeleton size 19-34 µm spines.

Class: Prymnesiophyceae (Haptophyceae) (Coccolithophorids)

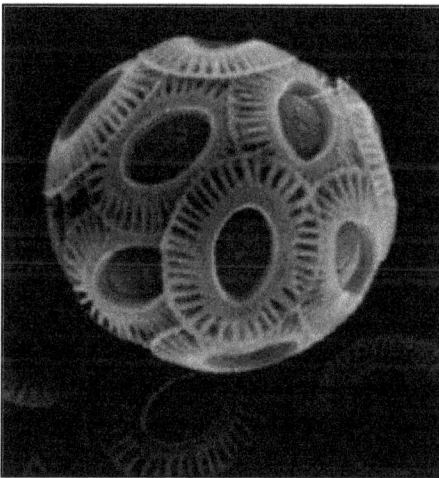

Cell spherical, round or flattened, elongated or saddle-shaped, with two flagella. Cell body surface covers with organic scales and large spiny scales. It has the haptonema, short or long thread-like organelle may be protruding in the swimming direction. Long haptonema coil when relaxed; yellow-brown to golden-brown, one or two chloroplasts.

Class: Euglenophyceae

Flagella-one, two, or four emergent, running from their bases in the reservoir through the canal. Non-emergent flagellum to the level of the eyespot. Mode of swimming-homo- or heterodynamic. Color-bright green in phototrophic forms. Chloroplastsone or many, reticulated, ribbon or disc shaped. Pyrenoids-often with paramylon shields or clusters of paramylon granules. Eyespot-orange or red, usually conspicuous, situated near the canal plasmalemma, separate from the chloroplasts. Nucleus-large, with condensed chromosomes, often prominent, in the middle of posterior part or the cell. A contractile vacuole is lacking in true marine species.

Class: Prasinophyceae

Cell shape-quadrangular or bilaterally compressed, often with a depression where the flagella originate. Cell covering-organic scales cover cell body and flagella, which may assemble to form a theca (*e.g.,* Tetraselmis). Naked species also occur. Flagella-one, two, four, eight (or 16) covered with minute scales and simple hairs, appear rather stiff and "thick". Mode of swimming-hetero- or homodynamic, flagella pushing the cell. Color-slightly olive-green. Chloroplasts-one (or two) simple or lobed campanulate, or many disc-shaped (phycoma stages). Storage product-starch in shield around pyrenoid, and as stroma starch in the chloroplast.

Pterosperma sp.

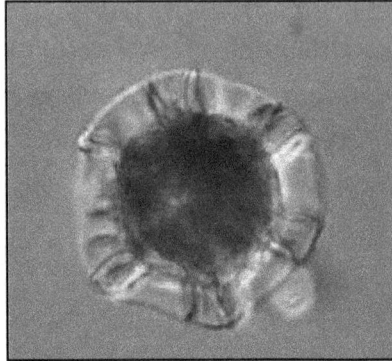

Motile cells bilaterally symmetrical with four long flagella, nonmotile stage walled, with wings (algae), and many discoid greenish yellow to golden brown chloroplasts.

Class: Chlorophyceae (Green Algae)

Cell is rounded or ovoid, may be lobed, naked or with cellulose wall. One, two, four (or eight) flagella color bright green. One chloroplast, parietal or campanulate, lobed or reticulated with starch shield. Found in coastal waters near and on shore, rock pools.

Dunaliella salina (Dunal) Teodoresco 1905

Description: Free-swimming, unicellular, naked biflagellate of microscopic size, ellipsoidal or ovoid in shape with cup-shaped green chloroplast at the posterior end

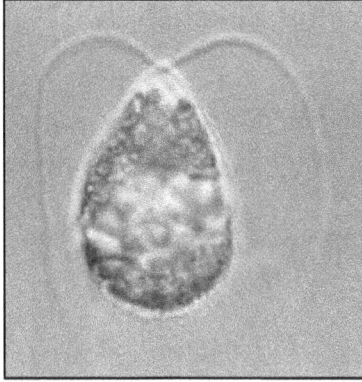

of the cell and two flagella. A girdle of refractive granules, which may be have a reddish tinge.

Dimensions: Cell length 16-24 µm.

Class: Cyanophyceae (Cyanobacteria or Blue-green algae)

Cyanobacteria have an algal-like morphology and unlike other bacteria, they perform photosynthesis. They are therefore classified as the class Cyanophysceae or the blue-green algae. However, they are purplish red, cellular plants; cells usually small. They lack cell organelles such as a nucleus, mitochondria and the chloroplasts and thus plastids are absent and colour is diffused throughout cell contents. Individual species vary from microscopic colonies of cells to elaborate filaments and are extremely common, especially on tropical shores.

Trichodesmium erythraeum Ehrenberg 1830

Synonym: Oscillatoria erythrea

Description: Colonies consist of straight trichomes with a parallel orientation. Ends of the trichomes may have a cap or calyptra. Central part of the cells has a granular appearance. Cells longer or shorter than wide.

Dimension: 7-12 µm wide, 60-750 µm long.

10.4. Toxic and Harmful Species

Toxic Species

Diarrhetic Shellfish Poisoning (DSP)

Causative organisms: *Dinophysis acuta, Dinophysis acuminata* and *Prorocentrum lima* (Dinophyceae).

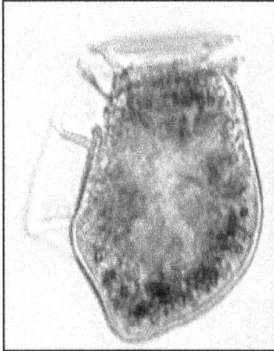

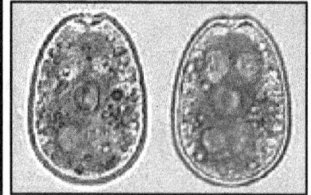

Dinophysis acuta   **Dinophysis acuminata**   **Prorocentrum lima**

Size range: Dinophysis acuta, 54-94µm in length, 43-60µm in width.

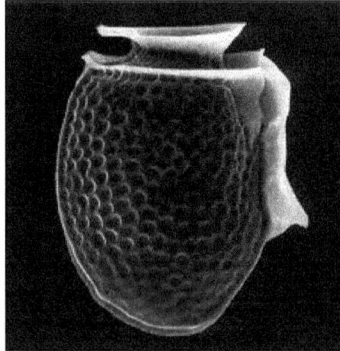

Dinophysis acuminata, 38-58µm long, 30-40µm width

Toxins: Okadaic acid (OA) and Dinophysis toxins (DTXs).

Symptoms: Diarrhoea, nausea, vomiting.

Problem: Accumulates in shellfish flesh and produces the toxins.

Paralytic Shellfish Poisoning (PSP)

Causative organisms: *Alexandrium minutum* and *Alexandrium tamarense* (Dinophyceae).

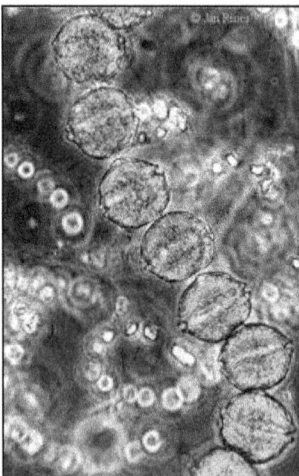

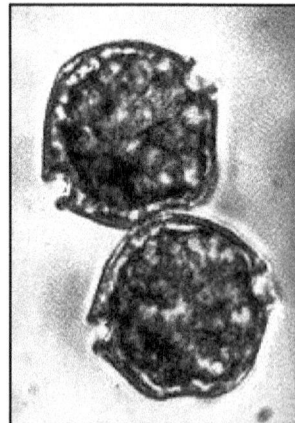

Alexandrium minutum   **Alexandrium tamarense**

Description: Cells are small and vary in shape; rounded or elongated.

Size range: *Alexandrium minutum,*15-30µm in length, 13-24µm in diameter.

Alexandrium tamarense, 22-51µm in length, 17-44µm in width.

Toxins: Saxitoxins, Gonyautoxins

Symptoms: Headaches, dizziness, diarrhoea, nausea, vomiting leading to muscular paralysis. In severe intoxifications respiratory failure may occur.

Problem: Accumulates in shellfish flesh and produces saxitoxins.

Alexandrium catenella

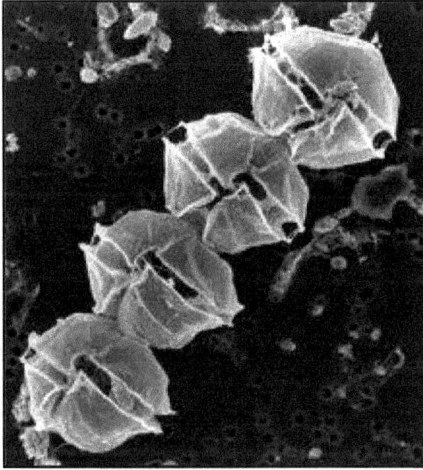

It is an armoured, marine, planktonic dinoflagellate. It is associated with toxic PSP blooms, especially in cold water coastal regions.

Taxonomic Description: It is a chain-forming species occurring in characteristic short chains of 2, 4 or 8 cells. Single cells are round, slightly wider than long, and are anterio-posteriorly compressed. A small to medium sized species, it has a rounded apex and a slightly concave antapex. Cells range in size between 20-50 µm in length and 18-30 µm in width.

Toxicity: *Alexandrium catenella* produces strong PSP toxins c1-c4 toxins, saxitoxins (SXT) and gonyautoxins (GTX) which are transmitted via tainted shellfish. These toxins can affect humans, other mammals, fish and birds.

Gymnodinium catenatum (Dinophyceae)

It is responsible for the red tides and is known to cause Paralytic Shellfish Poisoning (PSP). *G. catenatum* produces saxitoxin, which is a neurotoxin that disables the sodium pump. Mussels feed upon dinoflagellates such as *G. catenatum*. The toxin accumulates and concentrates in the filter feeders and humans ingest the shellfish. The symptoms for the PSP include a variety of gastrointestinal and neurological symptoms. Symptoms for a mild case include headache, nausea, vomiting, diarrhoea,

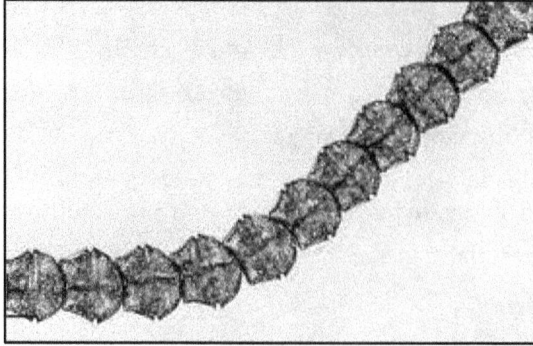

and a tingling sensation around the lips, gradually spreading to the face and neck. However, in an extreme case of PSP, the symptoms are much more severe. A person may experience muscle paralysis, respiratory difficulty, and death may occur 2-24 hours after ingestion.

Pyrodinium bahamense

Pyrodinium bahamense occurs throughout the world, though it is more common in the northern hemisphere. Cells are spherical with armored plates. When agitated, it responds by glowing. It has one horizontal spanning flagellum and one vertical propeller like one for locomotion. In large numbers, these dinoflagellates cause red tides. It produces neurotoxins that are especially potent. When conditions become unfavorable, the cells become a cyst and stay like that until conditions improve.

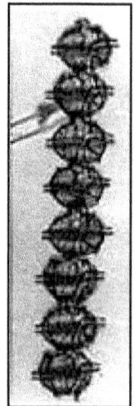

Toxicity: It is known for secreting toxins that cause Paralytic Shellfish Poisoning (PSP). The groups of toxins that are released by Pyrodinium bahamense are known as saxitoxins, which are sequestered in molluscian shellfish and are dangerous to mollusk consumers. Symptoms rapidly show up within an hour of eating contaminated shellfish, and consist of numbness, loss of motor function, incoherent, drowsiness and in the worst cases, respiratory paralysis.

Pyrodinium minimum

It is an armoured, marine, planktonic, bloom-forming dinoflagellate. It is a toxic cosmopolitan species common in cold temperate brackish waters and tropical seas.

Taxonomic Description: *Pyrodinium minimum* is a bivalvate species often seen in valve view. Cells are small (14-22 μm long to 10-15 μm wide) and shape is variable. Cells range from triangular to oval to heart-shaped. Cells are laterally flattened. A short apical spine is sometimes observable. Valves with short, evenly shaped broad-based spines arranged in a regular pattern. These can appear as rounded papillae depending on angle of view.

Toxicity: *Pyrodinium minimum* is a toxic species. It produces venerupin (hepatotoxin) which has caused shellfish poisoning resulting in gastrointestinal illnesses in humans and a number of deaths. This species is also responsible for shellfish kills in many countries.

Rhizosolenia chunii

Bloom of the diatom *Rhizosolenia chunii* may develop a bitter taste in mussels. The bitter taste is concentrated in the digestive gland, which may show extensive inflammation and degeneration. This species is the first species of diatom to be associated with shellfish mortality and, as the mortality has been reported to occur many months after the bloom had ceased.

Azaspiracid Poisoning (AZP)

Protoperidinium crassipes

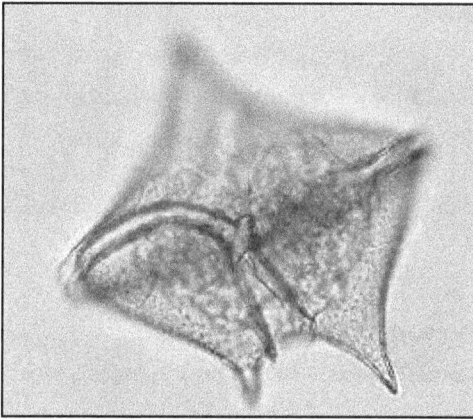

Body low and stout; slightly compressed dorsoventrally; ventral side rather concave but dorsal convex; apical horn conical abtuse; antapicals short, stout and close together; antapicals end in a blunt, semi-truncated projection with 2-3 points; right antapical longer than left; 70-85 by 60-72μm.

Causative organisms: *Protoperidinium curtipes, P.oblongum, P. Brevipes* and *P. stenii.*

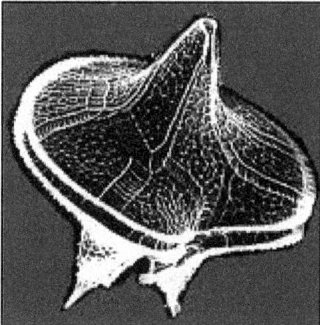

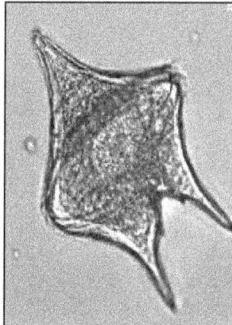

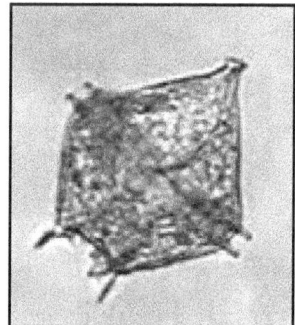

Protoperidinium curtipes   *P. oblongum*   *P. brevipes*

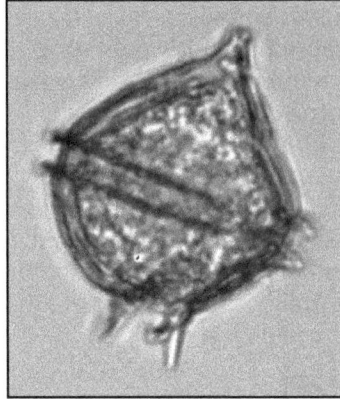

P. stenii

Toxins: Azaspiracids.

Symptoms: Same as DSP (more acute) with headaches and chills.

Problem: Azaspiracid toxins accumulates in shellfish flesh.

Description: These groups of organism are known as armoured, thecate dinoflagellates. Cells are characterized by two antapical horns or spines and one apical horn. Shape varies between species. These are single cell organisms and donot usually form chains.

Size range: 20-150μm in length.

Amnesic Shellfish Poisoning (ASP)

Causative organisms: Pseudo-nitzschia seriata and *P.delicatissima.*

Toxins: Domoic acid.

Symptoms: Diarrhoea, nausea, vomiting, abdominal pain, short-term memory loss.

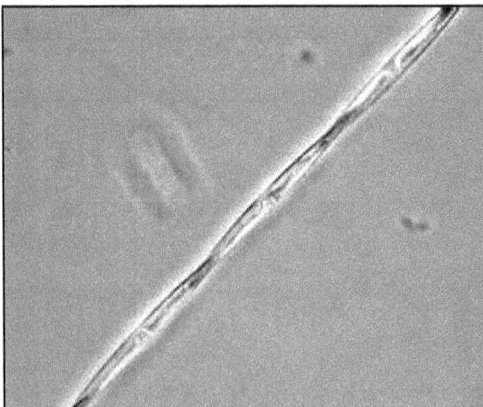

Pseudo-nitzschia seriata

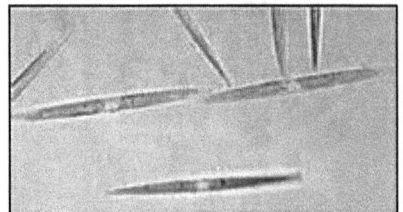

P.delicatissima

Problem: Domoic acid accumulates in shellfish flesh.

Description: Cells are elongate tapering at both ends. Joined in stepped chains by overlapping of cell ends. Distributed widely in many coastal areas of the world.

Size range: *Pseudo-nitzschia seriata*- more than 3µm in width.

P.delicatissima -less than 3µm in width.

Homo-yessotoxin

Causative organisms: *Protoceratium reticulatum* and *Lingulodinium polyedrum.*

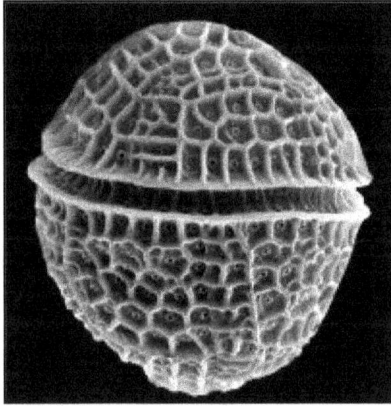

Protoceratium reticulatum   **Lingulodinium polyedrum**

Protoceratium reticulatum

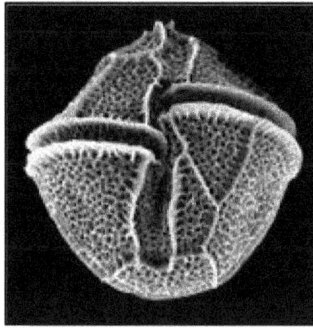

Symptoms: Can cause liver and heart damage.

Problem: Accumulates in shellfish flesh. Related to DSP toxins.

Description: *Protoceratium reticulatum* are small cells polyhedral shaped with strong reticulations. Lingulodinium polyedrum are polyhedral shaped with no spines or horns.

Size range: *Protoceratium reticulatum*: 28-43µm long and 25-35µm wide.

Lingulodinium polyedrum: 42-54µm wide.

Pectenotoxin (PTX)

Causative organisms: *Dinophysis fortii.*

Symptoms: Can cause liver and heart damage.

Problem: Accumulates in shellfish flesh. Related to DSP toxins can produce toxins in relatively low numbers.

Description: It is a medium size cell, broadly sub-ovoid, broadest posteriorly. Sulcal lists are well developed. Red- brown chloroplasts are present widely distributed from cold temperate to tropical seas.

Size range: 62-66µm long.

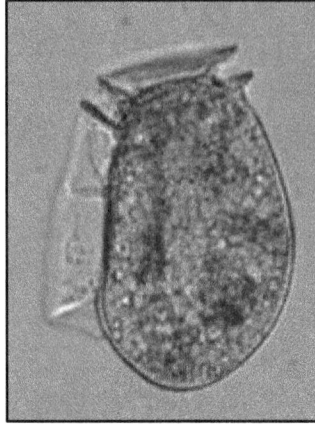

Dinophysis fortii

Ciguatoxin

Gambierdiscus toxicus is an armoured, toxic benthic dinoflagellate species attached to the surface of brown macroalgae. Cells of *G. toxicus* are are frequently found as epiphytes on macroalgae and dead corals.

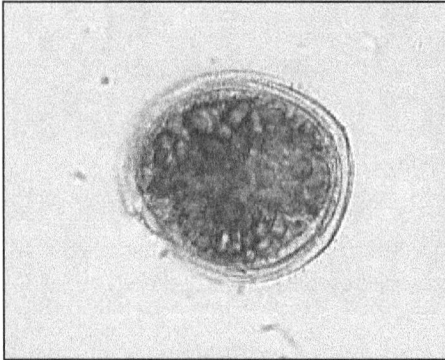

Gambierdiscus toxicus (Ventral)   *Gambierdiscus toxicus* (Lateral)

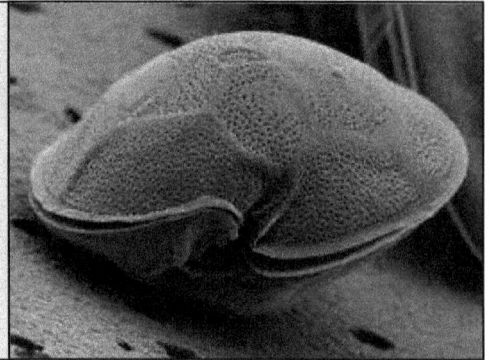

Taxonomic Description: Cells of *Gambierdiscus toxicus* are large, round to ellipsoid and flattened anterio-posteriorly. The epitheca and hypotheca are nearly equal in height. The cell surface is smooth with numerous deep and dense pores. Cells range in size from 24-60 µm in length, 42-140 µm in transdiameter, and 45-150 µm in dorso-ventral depth.

Toxicity: *G. toxicus* is known to produce toxins such as ciguatoxin, gambieric acid and maitotoxin.

Palytoxin

Ostreopsis siamensis is an armoured, marine, benthic dinoflagellate species.

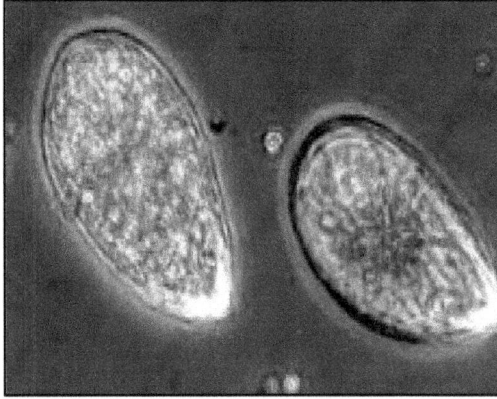

Ostreopsis siamensis

Taxonomic Description: Cells of *O. siamensis* are ovate and tear-shaped. The thecal surface is smooth with evenly scattered round pores. Large (0.5 um diameter) and small (0.1 um diameter) pores are present. Cells have a dorsoventral diameter of 108-125 um and a transdiameter of 76-85 um.

Toxicity: This species produces an analog of palytoxin.

Ichthyotoxic Species

Species of phytoplankton such as *Chaetoceros convolutus, Chaetoceros concavicornis, Chaetoceros densus, Chaetoceros eibenii* and *Chaetoceros danicus* are not toxic but when present in the water column in huge quantities can create problems to farmed finfish.

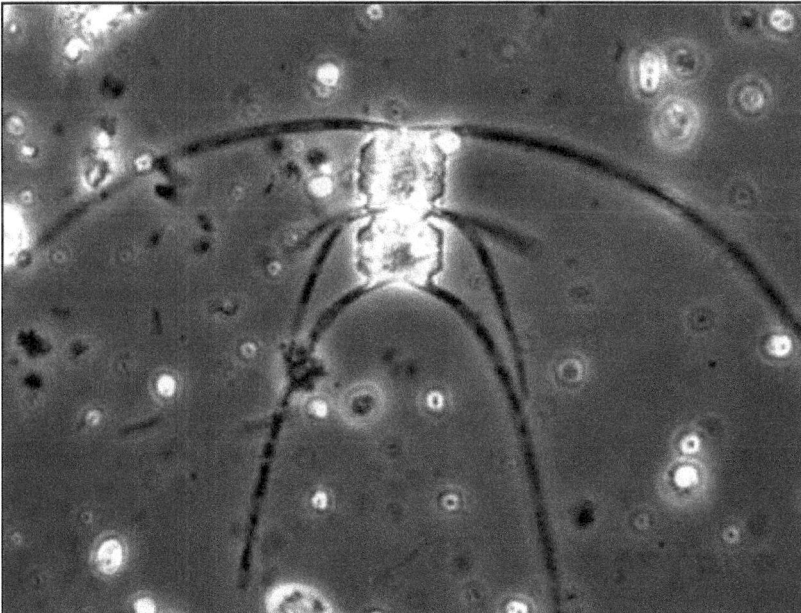

Chaetoceros convolutus

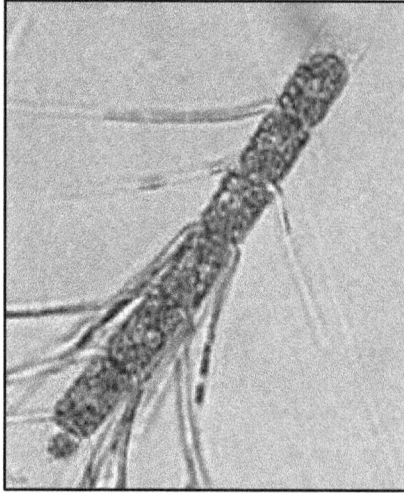

Chaetoceros concavicornis

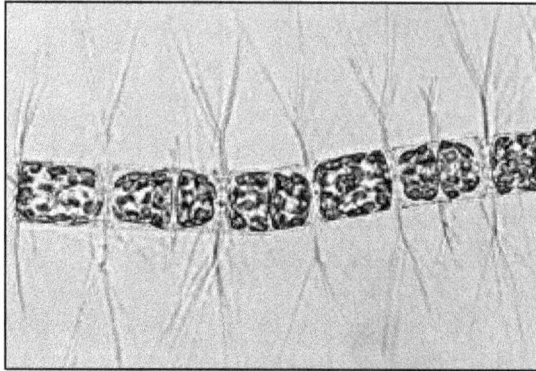

Chaetoceros densus

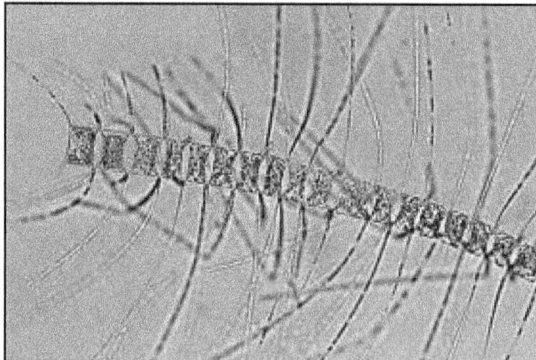

Chaetoceros eibenii

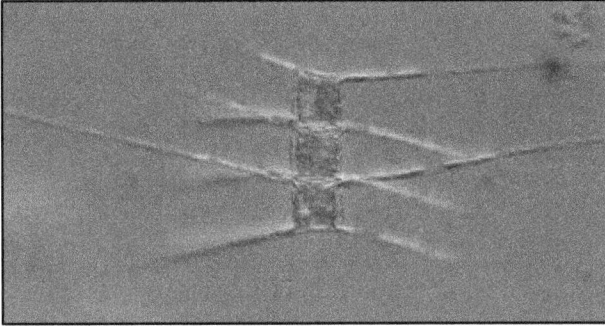

Chaetoceros danicus

These organisms when appear in numbers get contact with farmed fish and clog their gills, causing irritation. The resulting mucus production by the gill tissue stresses the fish, sometimes causing mortalities of farmed finfish.

Cochlodinium polykrikoides

It is an unarmoured, marine, planktonic dinoflagellate species with a distinctive spiral-shaped cingulum. It is a common red tide former associated with fish kills in many countries.

Taxonomic Description: *Cochlodinium polykrikoides* is an athecate species; *i.e.* without thecal plates. Cells are small, oval and slightly flattened dorso-ventrally. Chains, rarely more than eight cells, are common. Cells range in size from 30-40 um in length to 20-30 um in width. The epitheca is conical and rounded at the apex. The hypotheca is bilobed. The cingulum is deep and excavated. The narrow and shallow sulcus nearly runs parallel to the cingulum. The sulcus deepens and widens towards the antapex and divides the hypotheca into two asymmetrical lobes. Cells range in size from 30-40 um in length to 20-30 um in width. It is a common red tide former associated with fish kills in many countries.

Gonyaulax polygramma

Description: Cell elongated with epitheca exceeding hypotheca; Epitheca convex to angular, tapering to the apical horn; hypotheca symmetrical, rounded or truncate, may have a variable number of short antapical spines. Sulcus slightly excavated, widening to the posterior, and to the anterior running onto the epitheca. Plates ornamented with longitudinal ridges. Chloroplasts present; nucleus large, avoid, located in the posterior part of the cell.

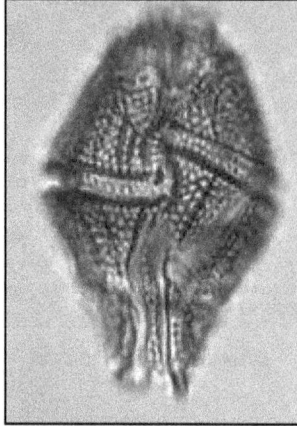

Dimensions: Length 29-66 μm; width 26-54 μm.

Prymnesium parvum (Golden algae)

This species has been reported to produce a number of toxins, collectively known as prymnesins, which include an ichthyotoxin, or fish toxin, a cytotoxin (a substance that is toxic to cells) and a hemolysin (a protein that causes the destruction of red blood cells). The ichthyotoxin adversely affects gill-breathing organisms of fish, bivalves, crayfish etc. The toxin damages the permeability of gill cells, which makes them susceptible to any toxins present in the water, including the *P. parvum* toxin itself.

The gills lose their ability to exchange water and absorb oxygen and bleed internally, resulting in death of the organism by asphyxiation.

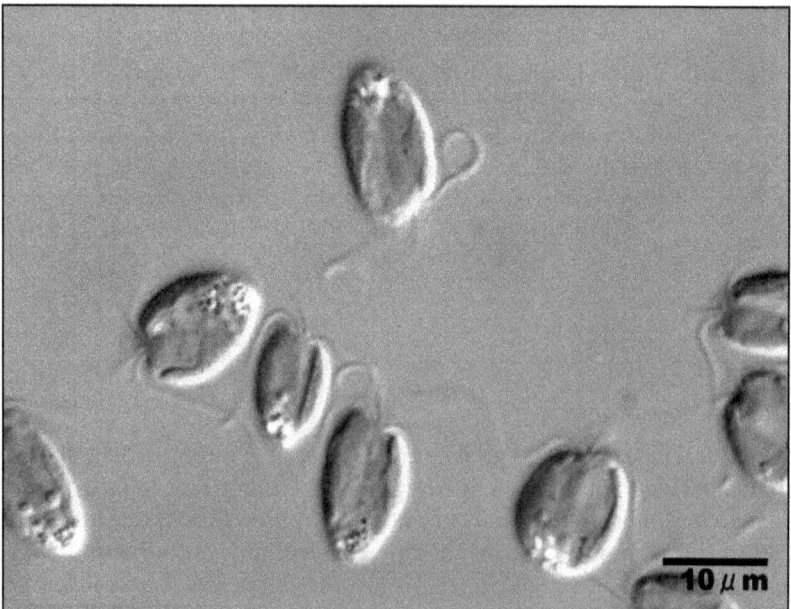

Harmful Species

Some phytoplankton species are not toxic, but can bloom in ideal conditions creating harmful algal events (HAE) or red tides which may cause fish kills and shellfish spat mortalities.

Noctiluca scintillans

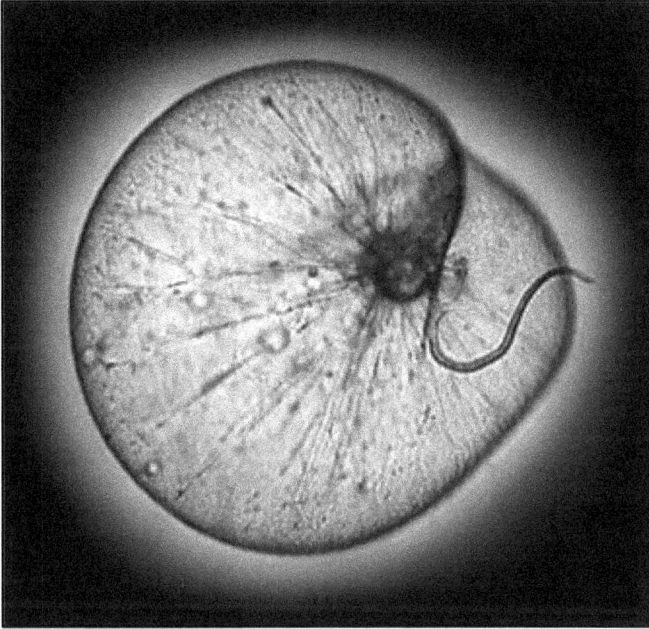

Synonym: *Noctiluca miliaris.*

Description: Body inflated, somewhat reniform; epicone and hypocone are almost same; sulcus very deep, mouth region extended anteriorly as an apical trough; longitudinal flagellum short, transverse flagellum as a mobile tooth; tentacle at posterior end of sulcus; Known for its large size and bioluminescent characteristics.

Dimension: Diameter 1,000 µm

Karenia mikimotoi

These single celled organisms are naked dinoflagellates which may reach high concentrations in the order of millions of cells per litre and are associated with causing fish, shellfish and benthic invertebrate mortalities. The resultant blooms are red in colour. The cells are round/oval in shape, typically 20-25µm in length and 20-25µm in width.

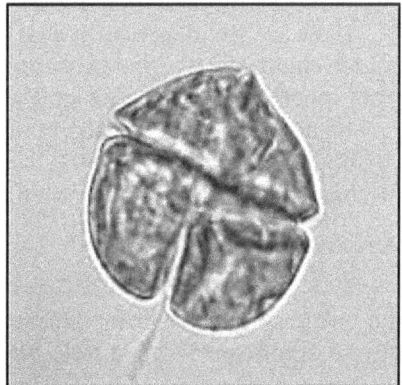

Phaeocystis sp.

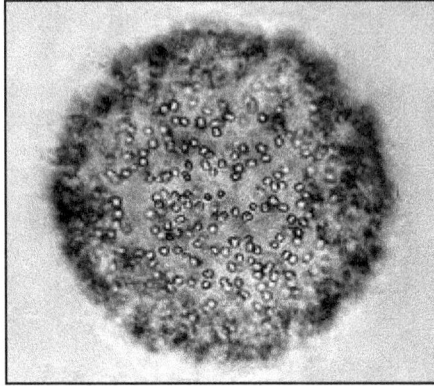

Cells are solitary or in gelatinous colonies which bloom very rapidly when conditions are favourable. They are not toxic, but they produce acrylic acid, dimethyl sulfide, and mucilage. Mucilage clogs gills of finfish and shellfish. Huge masses of foam are the characteristic of these blooms. These cells are cosmopolitan in distribution.

Heterosigma akashiwo

Cells are small, generally less than 25μm in diameter, and compressed. Many golden chloroplasts are present surrounding the cell periphery. Cell shape is variable from spherical to ovoid to oblong, sometimes appearing lumpy (potato shape). This organism can produce cysts, which under favourable conditions can germinate creating blooms. These blooms, which are brown in colour, are toxic to fish but the mechanism of toxicity is unknown. Cosmopolitan in distribution.

Amphidinium sp.

This genus of naked dinoflagellates is distinguished by its reduced epicone and the girdle situated near the anterior end of the cell. The species of this genus are littoral organisms living in rock pools, sand grains on beaches and salt marshes. If

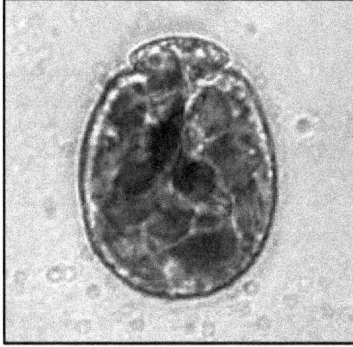

they bloom in sufficient numbers, they may discolour the beach green or brown at low tide. They are known to be involved in Ciguatera fish poisoning It has cosmopolitan distribution.

Chapter II
Identification of Zooplankton– Holoplankton

Holoplankton are organisms that are planktonic for their entire life cycle. Examples of holoplankton include radiolarians, forminifers, amphipods, krill, copepods, and salps. They normally live in the pelagic division as opposed to benthic division. The unique traits of holoplankton make reproduction in the water column possible. Both sexual and asexual reproductions are possible depending on the type of plankton. Some holoplanktonic invertebrates release sperm into the water column and females take up the sperm and fertilize the eggs. While others release the sperm and egg simultaneously in the planktonic environment in order to increase likelihood of fertilization. This release is triggered normally by environmental, mechanical, or chemical factors. Copepods are small holoplanktonic crustaceans that swim using their hind legs and antennae. Due to their small size and sluggish swimming abilities, holoplanktons have certain adaptations and are equipped with special defenses. Such adaptations include flat bodies, lateral spines, oil droplets, floats filled with gases, sheaths made of gel like substances, etc. Holoplanktonic zooplankton have also adapted by developing transparent bodies, bright colors, bad tastes and cyclomorphosis. Studies have shown that although small in size certain gelatinous zooplanktonare are rich in protein and lipid. Many holoplankton seem to have very little visible defense mechanisms and it is hypothesized that a chemical defense may be possible. Pelagic cnidarians have nematocysts that eject a tightly coiled venomous thread very rapidly. These threads penetrate the surface of their target and release venom.

11.1. Protozoans

Protozoans are the third, and last, common planktonic group that belongs to the Kingdom Protista. These single celled organisms are normally heterotrophs and do not contain photosynthetic pigments. Consequently they are mostly without colour and appear clear under the microscope. The phylum Protozoa is often divided into four classes: Ciliates, Flagellates, Amoebas and Sporozoans. Many of the Ciliates and Flagellates are extremely small members and belong to Nanozooplankton.

Planktonic Foraminiferans

Planktonic foraminiferans are unicellular organisms with a complex cell (Eukaryotes), and genetic material within a cell nucleus. They live floating in the surface waters of the open ocean, and secrete a calcium-carbonate shell. These shells make up the oozes forming on the oceans' floors.

The shells are commonly divided into chambers that are added during growth, though the simplest forms may be open tubes or hollow spheres. Fully grown foraminiferans range in size from about 100 micrometers to almost 20 centimeters long. Some have a symbiotic relationship with algae, which they develop inside their shells. Other species eat foods ranging from dissolved organic molecules, bacteria, diatoms and other single-celled algae, to small animals such as copepods. They catch their food with a network of thin pseudopodia (called reticulopodia) that extend from one or more apertures in the shell. There are about 4,000 species of foraminiferans living in the world's oceans and of them, 40 species are planktonic.

Phylum: Protozoa; Class:Sarcodina; Order: Foraminifera

Globoquadrina pseudofoliata

The test of this species is tightly coiled and low trochospiral with 3 to 3 globular chambers in the last whorl which increase rapidly in size and the last one is occupying up to one half of the entire whorl. The equatorial periphery is trilobate. The surface of the test has a distinct honeycomb pattern.

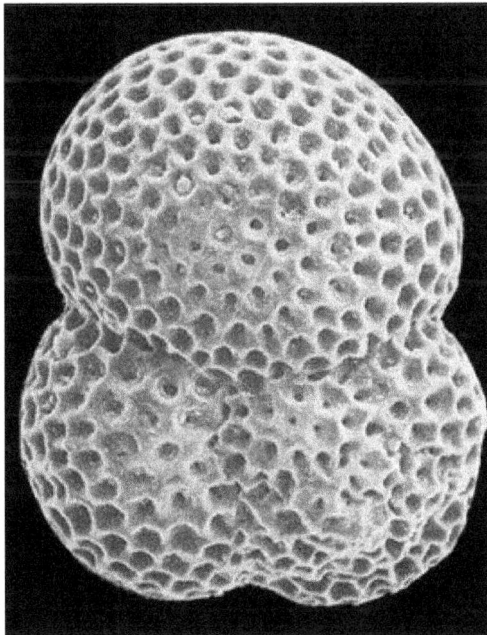

Globorotalia pseudobulloides

The shape of the test of this species is very low trochospiral, biconvex, and moderately compressed. Equatorial periphery is lobate. Axial periphery is rounded. The Wall is calcareous, perforate, surface smooth. Chambers are moderately compressed; 12 15, arranged in 2 2 1/2 whorls. The 5 chambers of the last whorl increase rapidly in size. Sutures on the spiral side are curved, less so in the last chambers, depressed and on umbilical side sutures are radial, depressed. Umbilicus is fairly narrow and open. Aperture is a low arch with a lip.

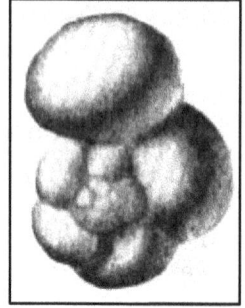

Radiolarians

The size of radiolarian species ranges from 30 microns to 2 mm in diameter. Their skeletons have arm-like extensions that resemble spikes, which are used both to increase surface area for buoyancy and to capture prey. Most radiolarians are planktonic. Most are somewhat spherical, but there exist a wide variety of shapes, including cone-like and tetrahedral forms. Besides their diversity of form, radiolarians also exhibit a wide variety of behaviors. They can reproduce sexually or asexually and are filter feeders or predators. They may even participate in symbiotic relations with unicellular algae.

Class: Actinopoda

Acanthometron sp.

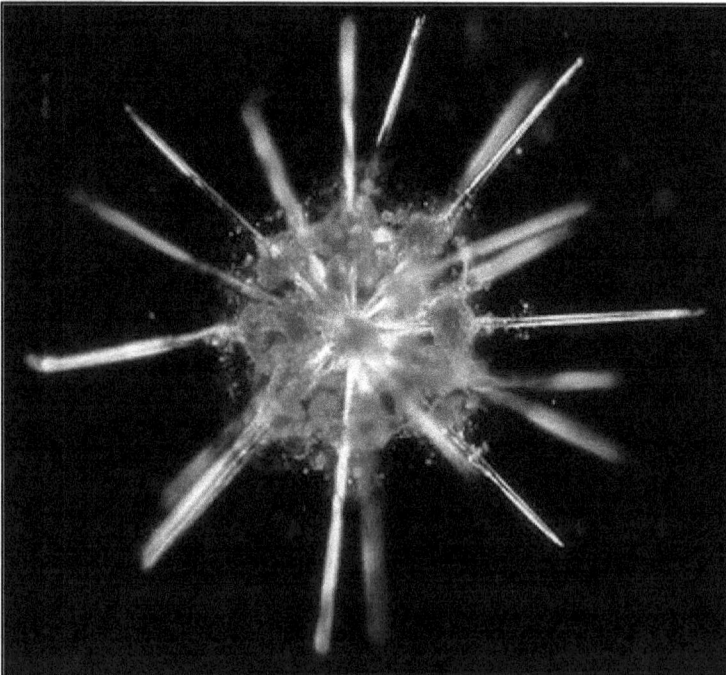

This unidentified species has a siliceous skeleton structure with radial symmetry and the skeleton spines are present at 360 degrees. It has a perforated central capsule of a horny substance which envelopes either a single very large nucleus or a number of smaller ones. Immediately outside this capsule, the protoplasm forms a dark layer in which digestion takes place.

Tintinnids

The tintinnids are ciliates distinguished by their vase-shaped shells called loricae. Like other protists, tintinnids are complex-celled (Eukaryota) organisms. They are heterotrophic and feed primarily on photosynthetic algae and bacteria. Majority of the tintinnids bear the size of microzooplankton (between 20 and 200 micrometres in size). Tintinnids are mainly found in marine, estuarine and brackish waters. Short generation times, high abundances, and fast reproduction rates, coupled with high grazing impact, signify the importance of this group as a key trophic link between the microbial and the metazoan compartment.

Tintinnopsis acuminata

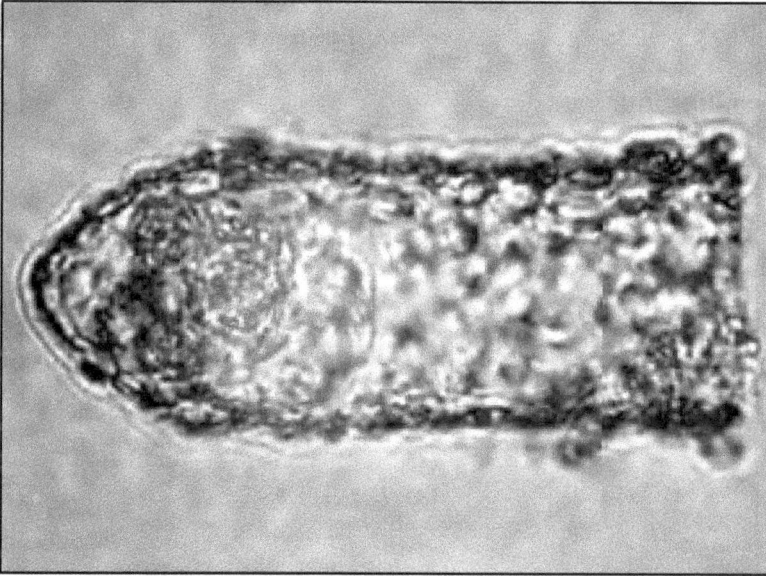

Description: Lorica tubular, oral rim ragged, aboral end blunt. Wall without spiral structure, with sparse agglomeration.

Measurements: Total length 54-78 µm. Oral diameter 18-21 µm. Approximate ratio L/oral diameter 2.9-3.5.

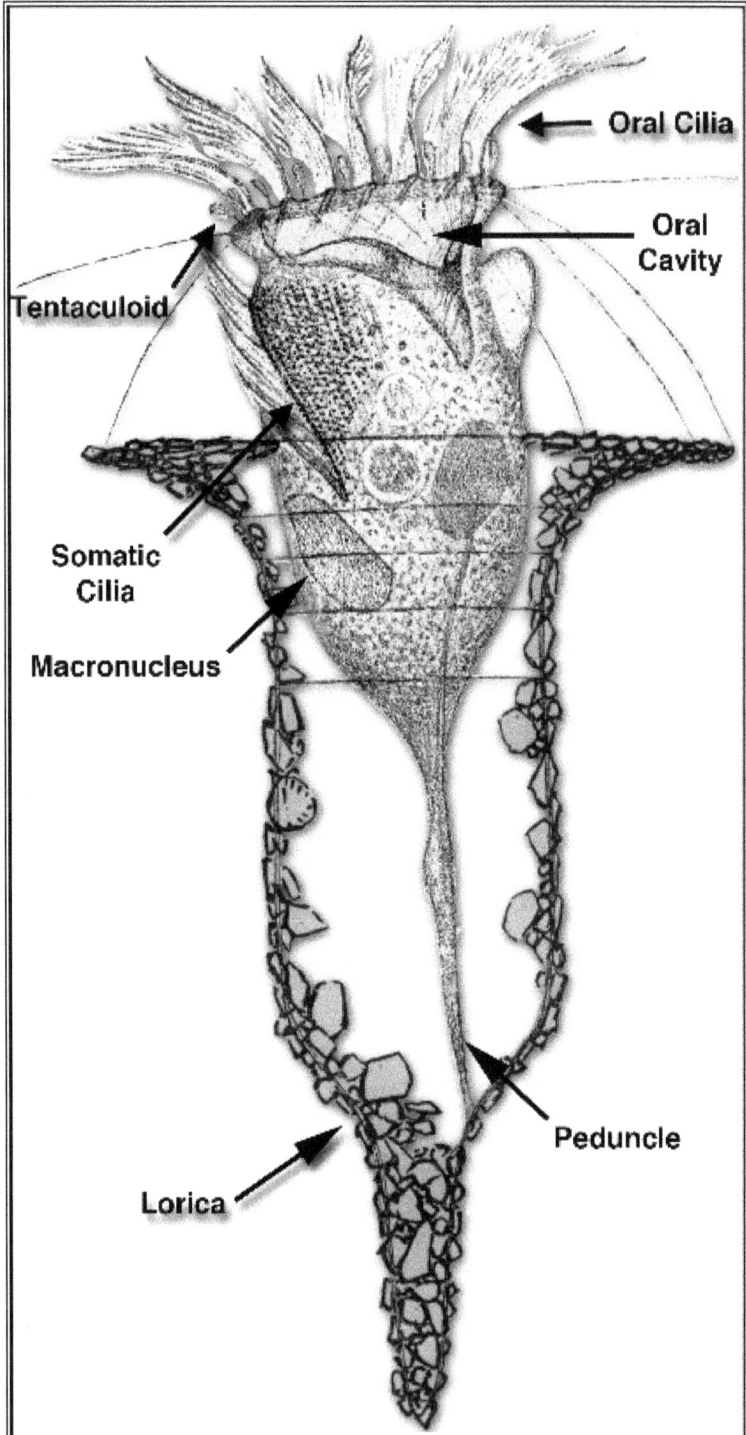

Internal Parts of a Typical Tintinnid Species.

Tintinnopsis ampla

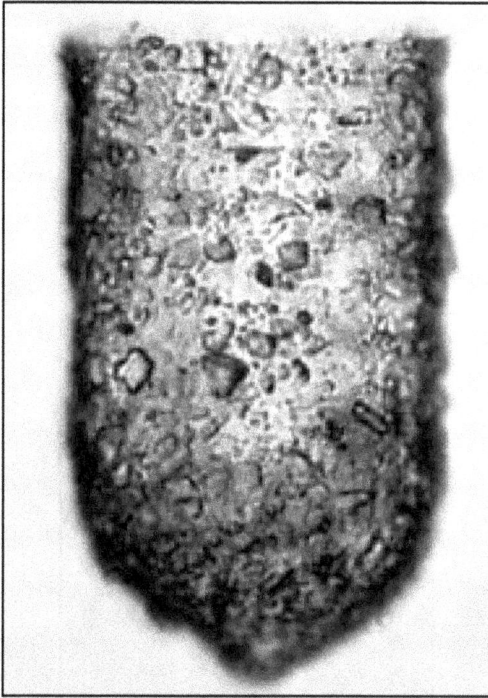

Description: Lorica comparatively large, generally bullet-shaped; ratio of lorica length to oral diameter is 1.8-2.6; oral margin entire; bowl cylindrical or very slightly tapering; aboral region tapering abruptly, rarely having a short aboral horn; aboral end blunt or acute; wall thickened, composed of rather coarse agglomerated materials, spiral structure visible, when well-developed sides of the bowl becoming uneven.

Measurements: Length 130-192 µm; oral diameter 70-75um.

Tintinnopsis beroidea

Description: Lorica bullet-shaped, usually cylindrical in the anterior 0.6 of the total length, aborally conical, oral rim ragged; aboral end acute or bluntly pointed; wall rather coarse, without spiral structure.

Measurements: Total length 34-110 µm. Oral diameter 18-40 µm. Approximate ratio of L/oral diameter 1.5-3.0.

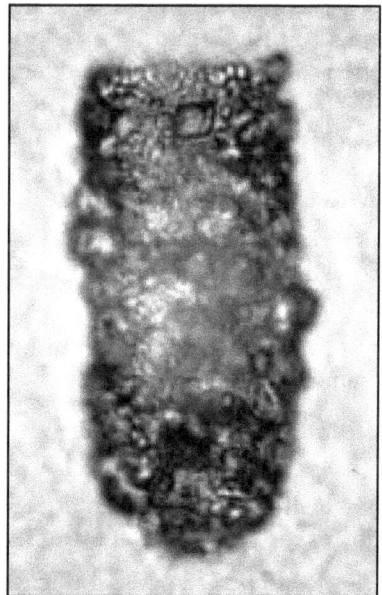

Tintinnopsis tocantinensis

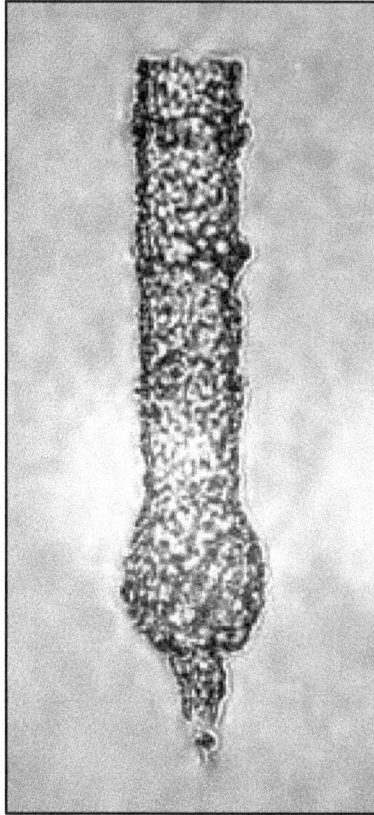

Description: Lorica elongated, cylindrical anteriorly, expanding posteriorly, tapering distally into a stout aboral horn; aboral horn conical, obliquely or irregularly open at the tip, wall thick and coarse.

Measurements: Total length, 103um. Oral diameter, 22um. Approximate ratio L/oral diameter, 4.7.

Tintinnopsis cylindrica

Description: Lorica is more cylindrical, elongated, without bowl but ending in an aboral horn. Agglomeration is similar to that of *T. tocantinensis.*

Measurements: Maximum total length, 540um and oral diameter 68um.

Tintinnopsis mortensenii

Description: This species is characterised by an agglomerate bowl which is enlarged anteriorly into a flare similar to that of a flower vase. Agglomeration is moderate on the bowl but dense on the rim of the flare.

Measurements: Maximum total length, 68 um and oral diameter 82um.

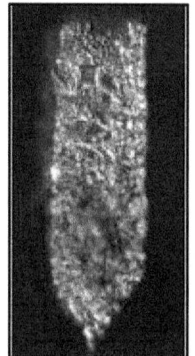

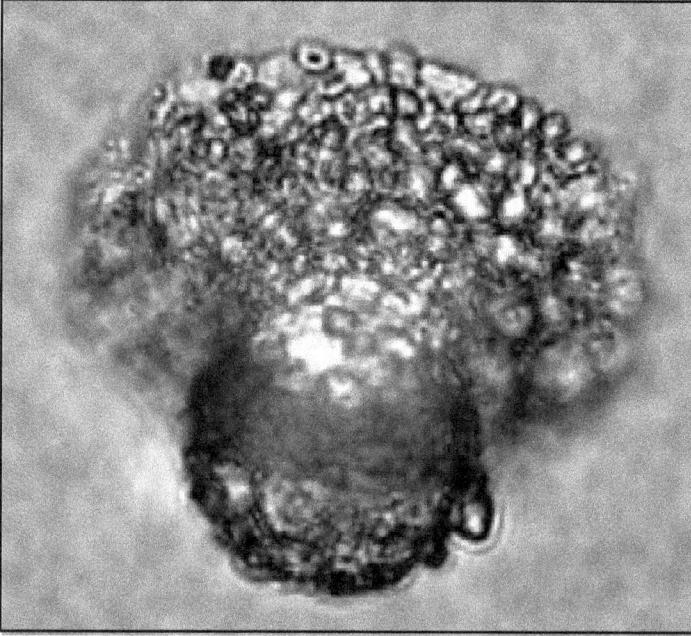

Tintinnopsis nordqvisti

Description: This species is characterised by an irregularly shaped posterior margin. Agglomeration is moderate.

Measurements: Maximum total length, 132 um and oral diameter 68um

Tintinnopsis tubulosa

Description: It is characterzsed by a cylindrical lorica ending with an indistinct bowl. Agglomeration is similar to that of *T.cylindrica.*

Measurements: Maximum total length, 165 um and oral diameter 68um.

Tintinnopsis directa

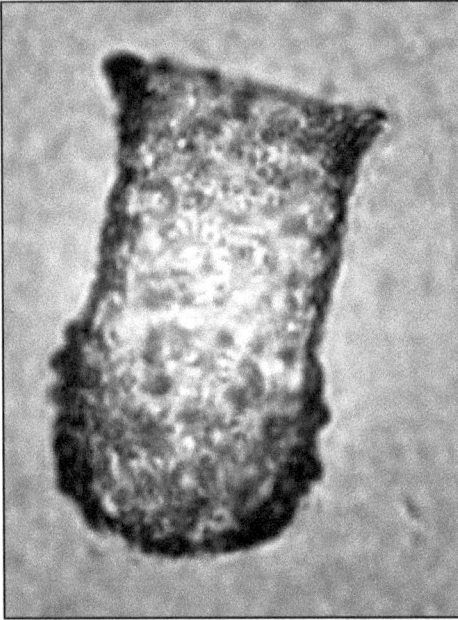

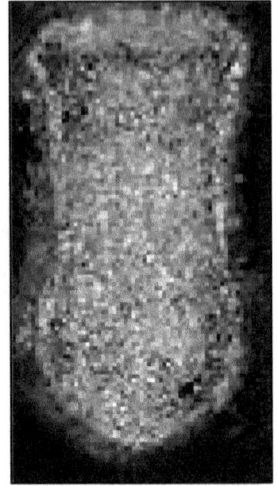

Description: Lorica tall, campanulate; oral rim irregular, flaring; suboral region somewhat apering, conical, laid up with about 6 spiral turns, narrowest at the basal portion of the subcylindrical part; posterior region subspherical, with a rounded aboral end; wall rather coarse in the posterior part.

Measurements: Total length 72-100 µm; oral diameter 35-48 µm; approximate ratio L/oral diameter 2.1-2.7.

Tintinnopsis radix

Description: Lorica elongate, slender, tubular, oral rim generally entire (smooth and round) or sometimes irregular; bowl long, cylindrical; aboral region tapering gradually into an aboral horn, inverted conical; aboral horn, usually more or less curved, with an irregularly formed aboral opening typically set laterally as gouged, leaving its tip or cutting it off; wall thin and fragile, with a variable spiral structure.

Measurements: Total length 140-524 µm. Oral diameter 32-75 µm. Approximate ratio L/oral diameter 6.0-9.5.

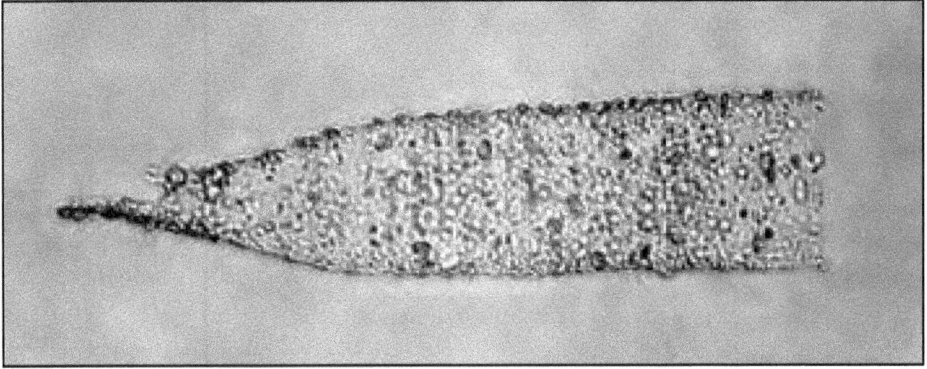

Tintinnopsis radix

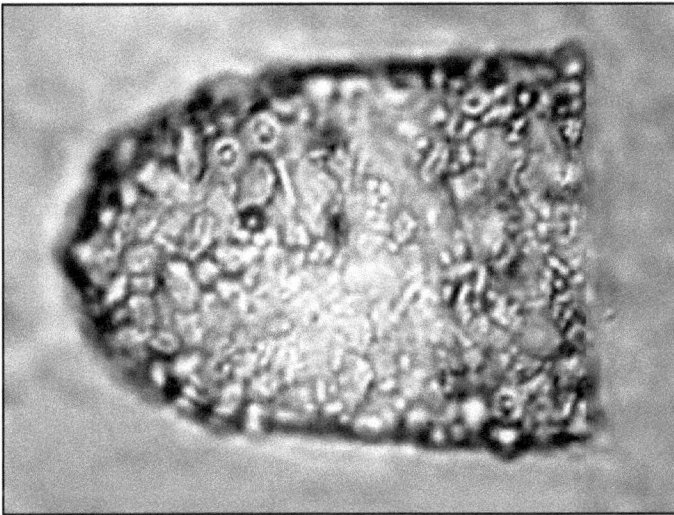

Tintinnopsis sacculus

Tintinnopsis sacculus

Description: Lorica short, cylindrical, with rounded aboral end; particles smaller and less numerous.

Measurements: Length 60-105 µm; oral diameter 44-58 µm; approximate ratio L/oraldiameter 1.5-2.0.

Tintinnopsis schotti

Description: Lorica bell-shaped, constricted in the suboral 1/4 of the total length; oral rim irregular; collar widely flaring to form an inverted truncated low cone with convex sides, its shortest basal diameter 0.68-0.78 of an oral diameter; bowl cup-shaped, usually broadest little below its middle; aboral region broadly convex conical to a blunt distal end; wall coarsely agglomerated, thickened to make an inward projection at the nuchal constriction, no spiral structure.

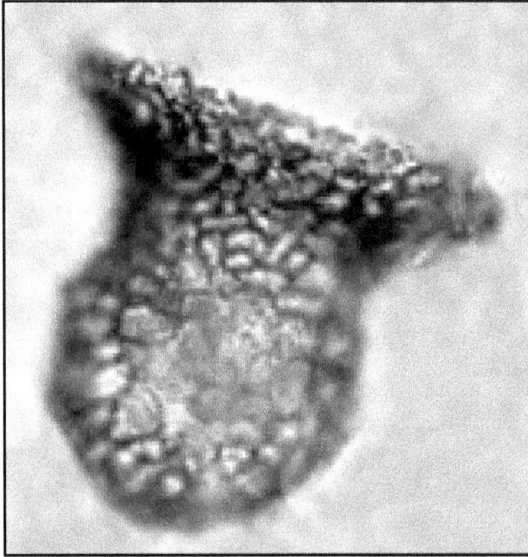

Measurements: Total length 96-120 μm. Oral diameter 82-85 μm. Nuchal diameter 58-65 μm. Greatest diameter of bowl 62-70 μm. Approximate ratio L/oral diameter 1.1-1.4.

Dictyocysta seshaiyai

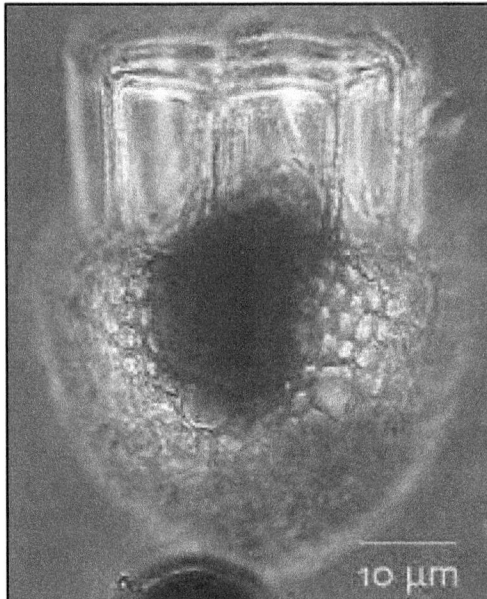

Description: This species was discovered and christened by the senior author. Lorica is divisible into bowl and collar portions. Collar possesses a single row of 4

rectangular windows formed by 4 frames placed at equidistance from each other. Bowl is bulbous but slightly pointed aborally. Agglomeration with foreign particles is noticed both on the bowl and collar; however, particles are larger in size in collar. Neither fenestrae nor reticulation is seen on lorica.

Measurements: Total length 72 to 83um. Inner oral diameter 49 to 54um.

Codonellopsis ostenfeldi

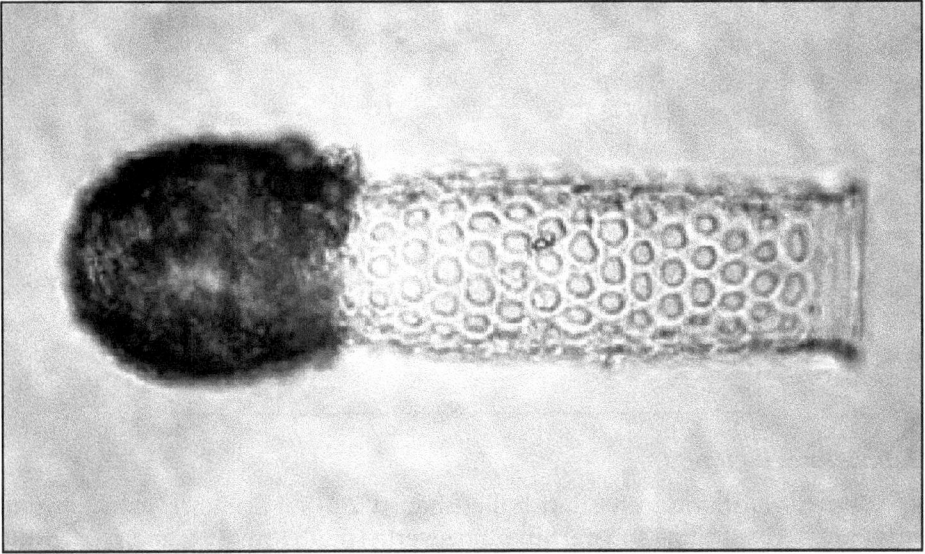

Description: Lorica flask-shaped, oral diameter is 0.37-0.33 of total length; oral margin entire (smooth and round), slightly flaring; collar subcylindrical, 0.40-0.43 of the total length in height, with 2-4 spiral turns in its anterior end and 5-6 annular rows of elliptical fenestrae lying transversely on the lower part; bowl globose, its greatest transdiameter 1.6 oral diameters; aboral end broadly rounded; wall agglomerated with coarse foreign particles.

Measurements: Length 95-112 µm; oral diameter 35-38 µm; length of the collar 38-50µm; greatest transdiameter of the bowl 55-62 µm.

Coxliella annulata

Description: Lorica tubular but very slightly wider toward aboral end, bluntly pointed aborally. Spiral turns slightly overlapping. Wall structure indistinct.

Measurements: Total length 269-332 µm; oral diameter 100-128 µm; approximate ratio L/oral diameter 2.7-3.0.

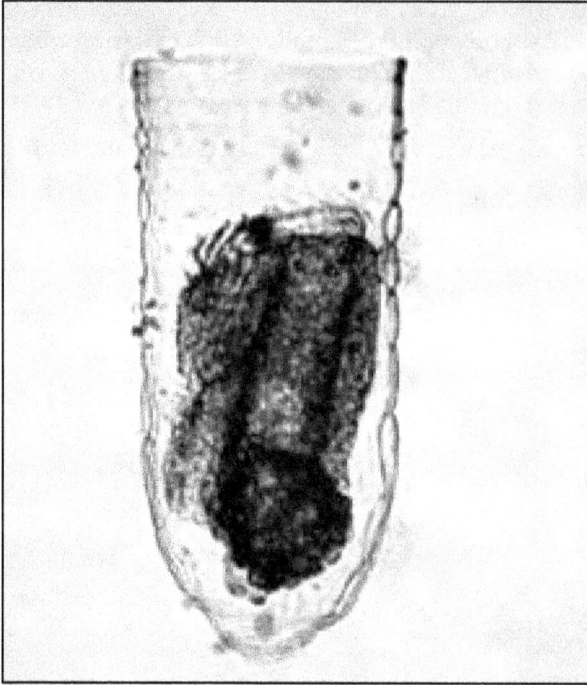

Helicostomella longa

Description: Lorica very short, bullet-shaped, 2-4 oral diameters in length; oral rim entire; 4-11 subequal suboral turns, not everted orally; bowl expanding slightly,

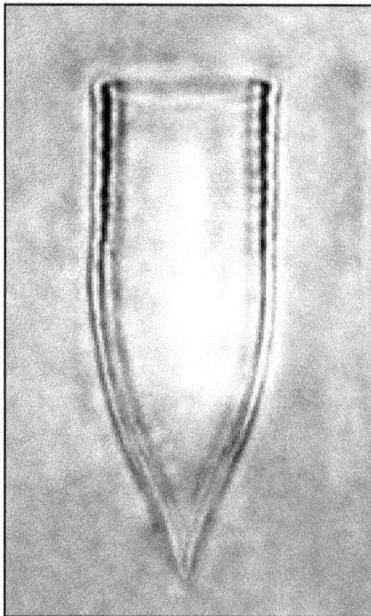

widest 0.5-1.0 oral diameter below the spiral, convex conical aborally; aboral horn scarcely differentiated, conical.

Measurements: Total length 50-80 μm.

Luminella sp.

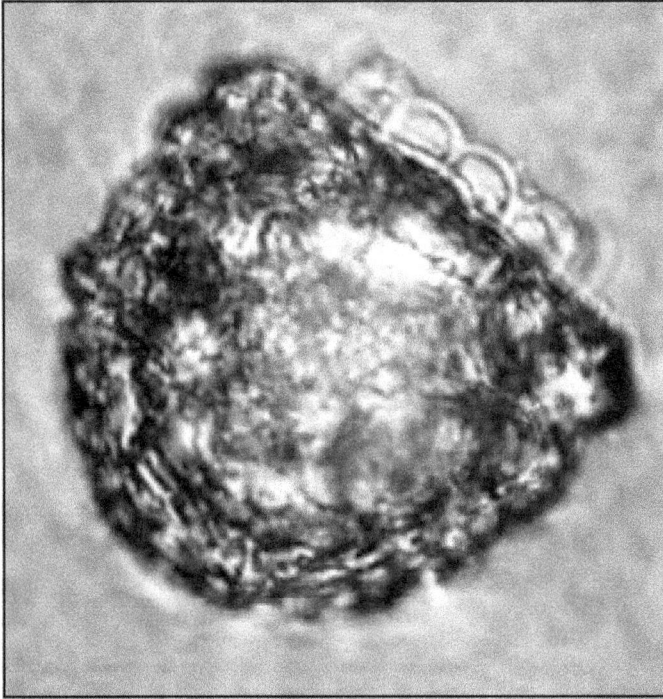

Description: Collar short, approximately 1/10 of total length, hyaline consists of 8-10 thin connected arcs enclosing big squarish fenestrations. Bowl cup-shaped, heart shaped or globose, bluntly pointed or hemispherical aborally, usually forms pronounced shoulder below collar. Surface densely covered with big agglomerated particles. Bowl is without any visible concentric structures.

Measurements: Total length 47.5-58.8 μm. Length of bowl 42.5-50.0 μm. Maximum width of bowl 42.5-52.5 μm. Diameter of oral aperture 22.0-32.5 μm. Length of collar 5.0-7.5 μm.

Metacylis jorgensenii

Description: Lorica short, ovoid, with slightly or sharply pointed aboral end. Collar short, with 2-5spiral turns, cylindrical or slightly flaring, narrower than bowl.

Measurements: Total length 50-61 μm; oral diameter 44-50 μm; approximate ratio L/oral diameter 1.3-1.9.

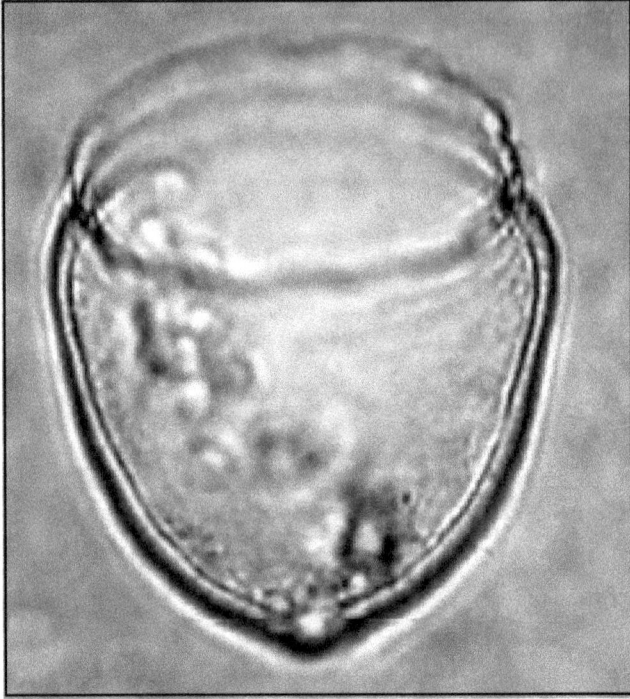

Metacylis tropica

Description: Lorica small, global, length is approximately equal to maximum width. Collar with 3-5 spiral turns, 0.30 of the total length of lorica.

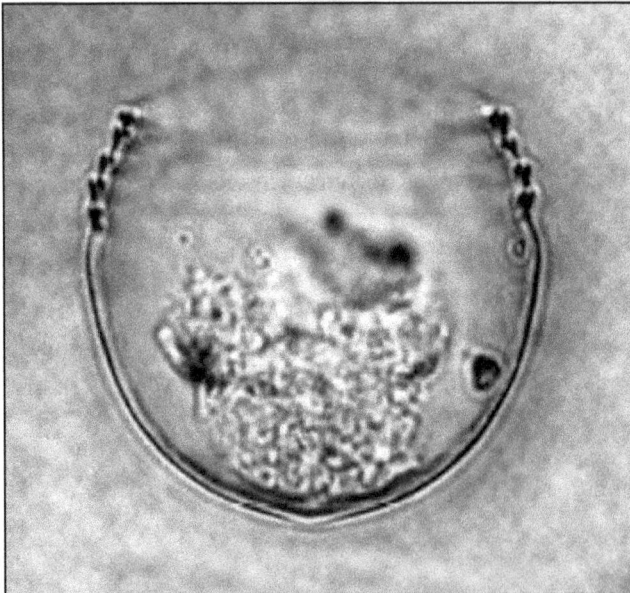

Measurements: Total length 43 μm; oral diameter 33-40 μm; maximum diameter 41-44 μm.

Rhabdonella conica

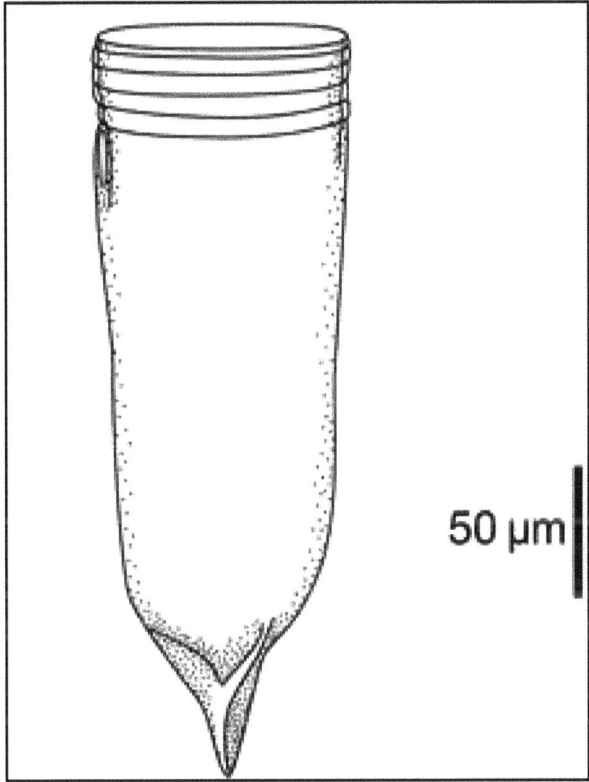

Description: Lorica very long, chalice-shaped, 7 oral diameters in length; oral rim flaring abruptly, no higher than lip; bowl tapering conical, it is about half the length of the total length of the whole lorica; pedicel very long, subcylindrical, tapering distally; aboral end open; ribs 32-48, with sinistral deflection, almost vertical on top half, slightly spiral to near aboral end; fenestrae distinct, widely distributed.

Measurements: Total length 290-470 μm. Oral diameter 37-102 μm. Approximate ratio L/oral diameter 4.8-7.9.

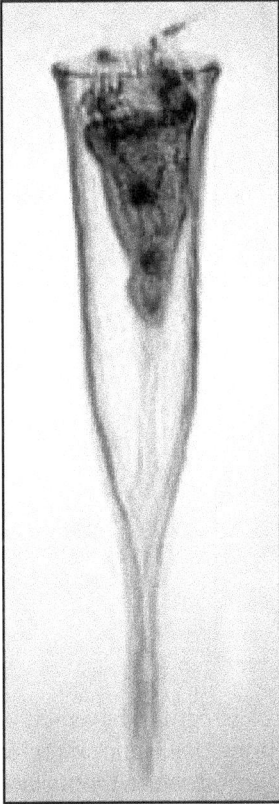

Rhabdonella conica   **Favella brevis**

Favella brevis

Description: This species possesses a fairly long and cylindrical lorica. The spiral lamina is formed of 4-5 turns which are wavy.

Measurements: Total length, 205um and oral diameter 103um.

Favella campanula

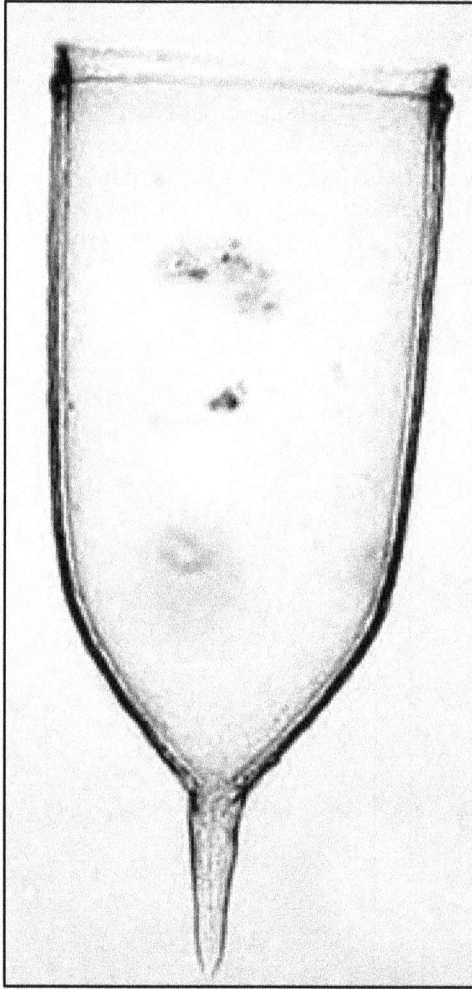

Description: Lorica campanulate, oral rim more or less irregular; bowl with a raised band just below the suboral ring, cylindrical in its upper 0.65; aboral region convex conical to a stout horn; aboral horn having a few vertical striae on the surface, tapering to a bluntly pointed tip, its length 0.15 of the total length; wall almost hyaline with a hardly visible reticulation, apparently separated in the suboral inflated part and scarcely in the following 0.15 of the total length.

Measurements: Length 143 µm; oral diameter 68 µm. Approximate ratio L/oral diameter 2.1.

Favella ehrenbergi

Description: Lorica long, cylindrical; bowl sometimes slightly expanded below middle, rounded below and joined by wings to a short blunt, pedicel. Spiral turns sometimes present suborally.Wall thick.

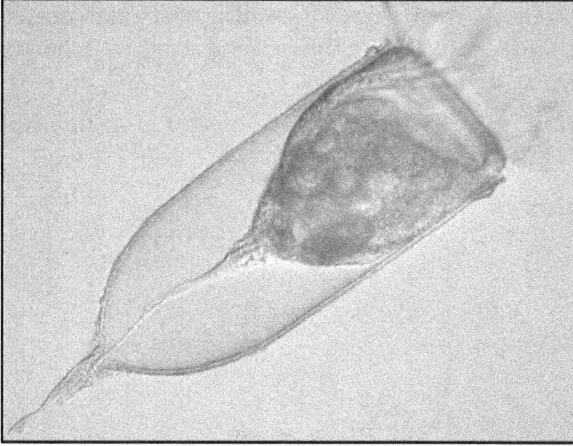

Measurements: Total length 145-400 μm. Oral diameter 54-124 μm. Approximate ratio L/oral diameter 2.4-4.2.

Amphorellopsis acuta

Description: Lorica fusiform, oral aperture circular; collar low-funnel-shaped, its nuchal smallest diameter 0.54-0.60 of an oral diameter; bowl circular in cross

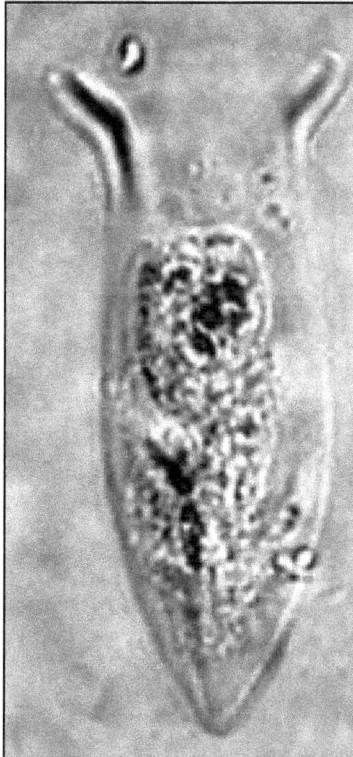

section below the collar, then gradually becoming triangular, posteriorly with three ridges in the aboral 0.6 of the total length; aboral end acute; wall composed of separated laminae in the anterior 0.2 of the lorica.

Measurements: Length 85-108 µm; oral diameter 31-45 µm. Approximate ratio L/oral diameter 2.3-3.9.

Amphorides amphora

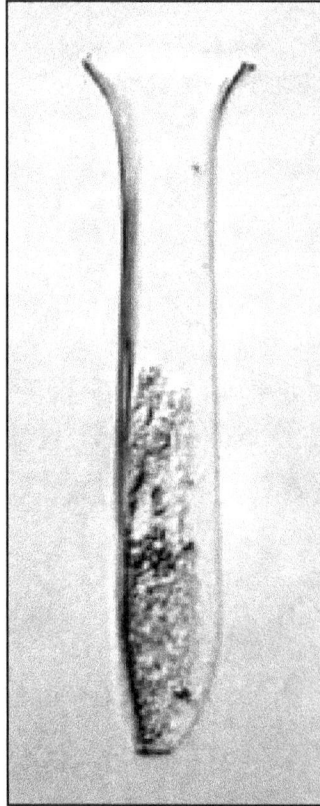

Description: Lorica with flaring collar; oral aperture circular; usually with flaring collar. Greatest width in lower part of bowl, with truncated aboral end. Three low vertical fins. Wall thickened at neck.

Measurements: Length 100-220 µm; oral diameter 55 µm, approximate ratio L/oral diameter 2.5-4.0.

Dadayella cuspis

Description: Lorica small, elongated; collar flaring and bowl extending to the pedicel; bowl subcylindrical conical posteriorly; pedicel short, 0.1-0.2 total length.

Measurements: Total length 67-93 µm; oral diameter 22-31 µm. Approximate ratio L/oral diameter 2.9-3.1.

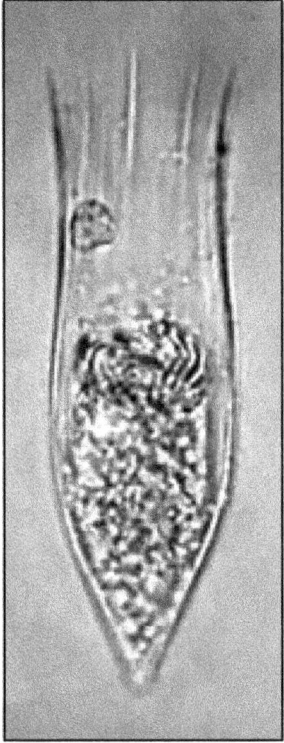

Dadayella cuspis   **Eutintinnus conicus**

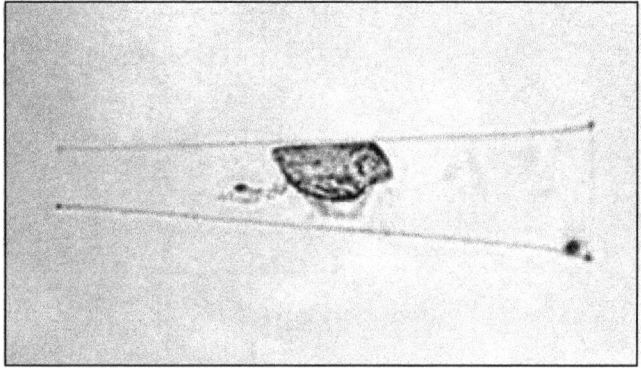

Eutintinnus conicus

Description: Lorica has a shape of a short truncated cone, narrowing.Oral end flaring to a slightly emergent brim. Aboral end strait. Aboral diameter little less than 1/2 oral diameter.

Measurements: Length 166-178 µm; oral diameter 42-48 µm; aboral diameter 21-23 µm; approximate ratio oral/aboral diameter 2.0-2.1; approximate ratio L/oral diameter 3.7.

Eutintinnus tenue

Description: Lorica almost cylindrical; with tapering shaft; oral end abruptly flaring into a horizontal brim; aboral end without flare.

Measurements: Length 179-238 µm; oral diameter 38-54 µm, approximate ratio L/oral diameter 4.3-6.1

Salpingella attenuata

Description: Lorica very much elongated, collar is a funnel, 0.6 oral diameter in length; bowl cylindrical in the anterior 0.8-0.9 total length, posteriorly a cone with

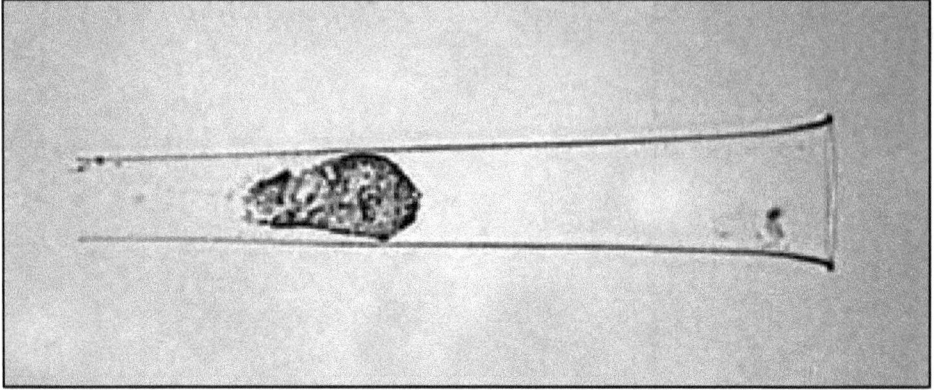

Eutintinnus tenue

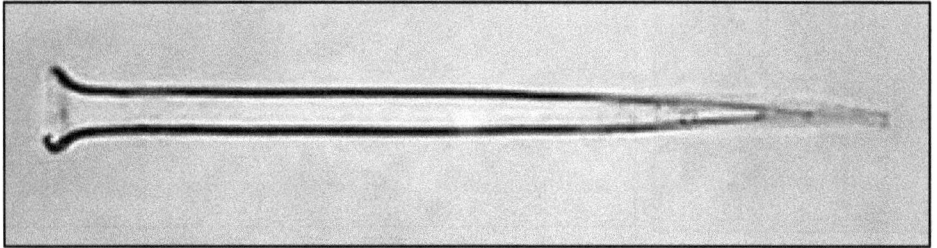

Salpingella attenuata

unevenly contracted sides; no aboral cylinder; fins 5-7, low blades, 0.2-0.4 total length in length, usually subvertical, sometimes giving an appearance of aboral expansion.

Measurements: Length 248-433 μm; oral diameter 32-43 μm, transdiameter of the bowl 16 μm, approximate ratio L/oral diameter 7.4-15.3.

11.2. Cnidarians (Coelenterates)

Planktonic cnidarians (Greek for "stinging nettle.") range in size from large jellyfish such as *Pelagia* sp. (over 2 meters long), to the microscopic medusa and larvae. Medusae represent one stage in the complex life cycle of a hydroid or jellyfish. All exhibit characteristic radial symmetry, *i.e.* they have a spherical or cylindrical body. They are at the tissue level of organization, not possessing true organs, and have only two kinds of tissues. Cnidarians are easy to spot with their ring of stinging tentacles surrounding a central mouth. This stinging capability makes this group mostly carnivorous.

Phylum: Cnidaria

Class: Hydrozoa

Order: Leptomedusae (Thecata)

Aequorea pensilis

Description: Leptomedusae with lens-shaped central disc. Tentacle bulbs with long lateral extensions, without excretory papillae. Umbrella up to 100 mm wide, more or less biconvex, with thin margin; manubrium 1/2 as wide as umbrella; gonads extending along almost entire length of manubrium; 150-250 radial canals; 10-16 tentacles and as many small rudimentary bulbs, no excretory papillae but excretory pores present as slits; statocysts many.

Eirene viridula

Description: Umbrella 30 mm wide; hemispherical, middle region fairly thick; with slender gastric peduncle with pyramidal base; manubrium rather small but with four long pointed lips with crenulated margins; with four radial canals; gonads linear, sometimes slightly sinuous, extending from somewhat beyond base of peduncle

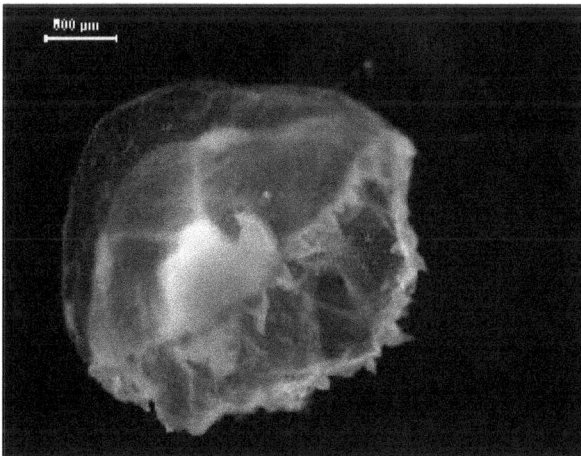

to almost bell margin about 70 tentacles of different sizes, large and small often alternating bulbs conical with distinct adaxial excretory papillae; 50 or more statocysts.

Order: Narcomedusae

Solmundella bitentaculata

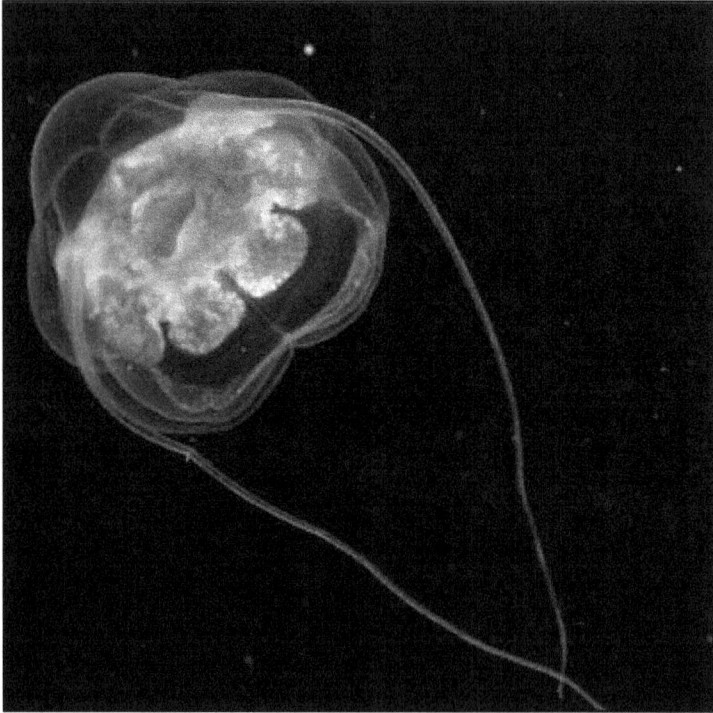

Description: This species has 8 manubrial pouches without peripheral canal system; only 2 long tentacles; without secondary tentacles; Umbrella 10 mm; apical mesoglea very thick; tentacles issuing from umbrella near apex, 8-30 statocysts.

Order: Trachymedusae

Liriope tetraphylla

Description: Umbrella 30 mm wide, hemispherical; manubrium on gastric peduncle of varying size; mouth with 4 lips; 1-3 (or more)centripetal canals in each quadrant; 4 long hollow perradial tentacles with cnidocyst rings and 4 small solid interradial tentacles with adaxial cnidocyst clusters; gonads variable in shape and size; 8 statocysts.

Order: Siphonophora

Diphyes dispar

Description: Nectosac of anterior nectophore cylindrical basally but with narrow caecal extension ending close to the apex of the nectophore. Hydroecium extends to

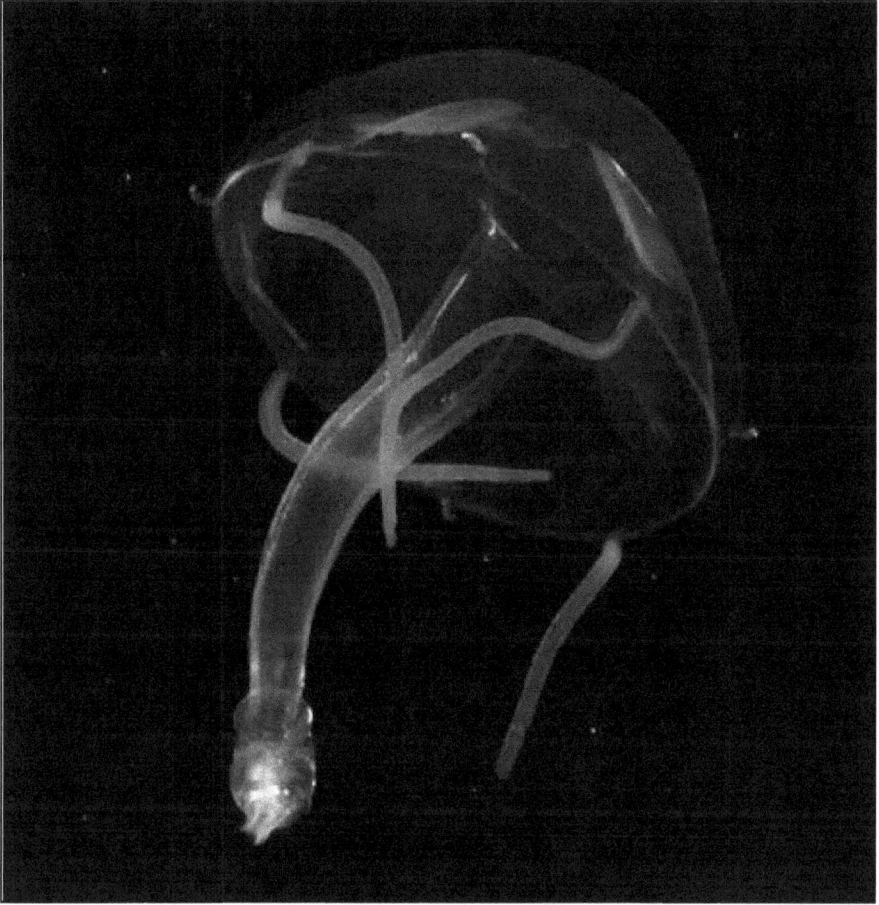

Liriope tetraphylia

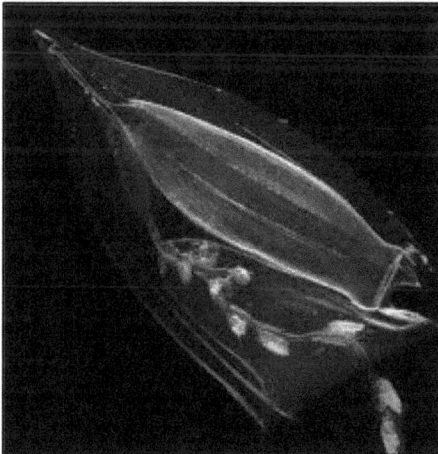

Diphyes dispar

about one-half the height of the nectophore, with the spindle-shaped somatocyst extending up to just above the beginning of the nectosacal caecum. Dorsal ostial tooth is considerably larger than the lateral ones. Mouth plate not divided.

Posterior nectophore with similar arrangement to ostial teeth. Lateral teeth and baso-lateral margins not serrated. Phyllocyst is narrower and tapers towards its apex, which lies further from the apex of the bract.

Physalia physalis

Description: The Portuguese man-of-war (*Physalia physalis*) is a floating hydrozoan colony, made up of four polyp types: pneumatophore (float), dactylozooids (tentacles for defense and prey capture), gastrozooids (feeding), and gonozooids (reproduction). They are recognized by the bluish pneumatophore, or float, which may be up to 30 cm in length. This float is an overgrown polyp that is oblong-shaped and filled with gas. In most siphonophorans, the gas is similar to the surrounding air composition, but in *Physalia* there is a greater concentration of carbon monoxide. The floats include a mechanism that controls the gas to regulate the depth of the organism, which in the case of men-of-war keeps them on the surface of the water. The tentacles, which can appear blue to purple, may reach lengths of up to 50 m. Men-of-war have two different sizes of stinging cells or nematocysts (small and large) for stunning or killing prey.

Porpita porpita

Description: *Porpita porpita*, commonly known as the blue button, is consisting of a colony of hydroids. The blue button lives on the surface of the sea and consists of two main parts: the float and the hydroid colony. The hard golden-brown float is round, almost flat, and about one inch wide. The hydroid colony, which can range from bright- blue- turquoise to yellow. Each strand has numerous branchlets, each of which ends in knobs of stinging cells called nematocysts. The blue button sting is not powerful but may cause irritation to human skin. The blue button itself is a passive drifter, and is part of the neustonic food web. It is preyed on by the sea slug *Glaucus atlanticus* (sea swallow or blue glaucus) and violet sea-snails of the genus *Janthina*. It competes with other drifters for food and mainly feeds on copepods and crustacean

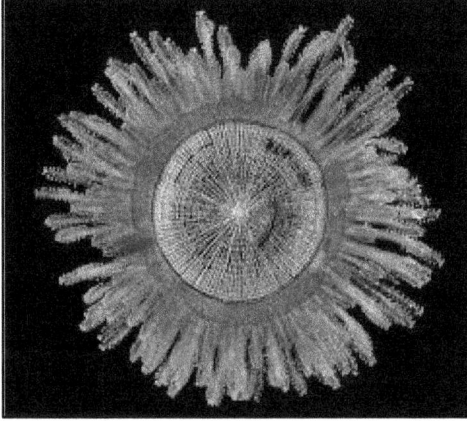

larvae. The blue button has a single mouth located beneath the float, which is used for both the intake of prey and for waste disposal.

Velella velella

Description: *Velella velella* is commonly known as sea raft, by-the-wind sailor, purple sail, little sail, or simply Velella. A type of jelly, the by-the-wind sailor sports a deep-blue, rectangular float topped with an upright, triangular sail. By-the-wind sailors often drift across the ocean's surface in large numbers, sometimes in the tens of thousands. During certain seasons, they are often blown ashore, blanketing coastal

beaches. These animals are made up of many individual polyps or zooids, making them a colonial animal. The zooids have different functions. The gastrozooids feed for the colony, using their tentacles to capture plankton. The gonozooids are the polyps that serve a reproductive function, constantly releasing tiny medusa (3 mm) that are the sexual reproductive stage of this animal.

Class: Scyphozoa

Order: Semaeostomeae

Aurelia aurita

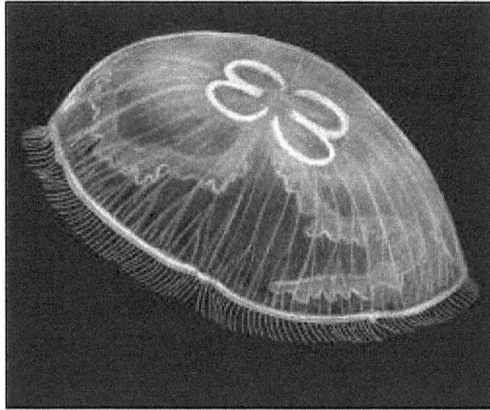

Description: *Aurelia aurita* is commonly called as the moon jelly, moon jellyfish, common jellyfish, or saucer jelly. The jellyfish is translucent, usually about 25–40 cm in diameter, and can be recognized by its four horseshoe-shaped gonads, easily seen through the top of the bell. It feeds by collecting medusae, plankton, and molluscs with its tentacles, and bringing them into its body for digestion. It is capable of only limited motion, and drifts with the current, even when swimming.

The body is a saucer shaped 'bell', which is colourless except for four obvious violet gonads visible in the centre of the disc. The outer edges of the bell are fringed with many small tentacles, and four stocky 'arms' surround the mouth.

Chrysoara colorata

Description: The bell of the purple-striped jelly (*Chrysaora colorata*, formerly *Pelagia colorata*) is up to 1m in diameter, typically with a radial pattern of stripes. When it is very young, it has a pinkish colour and its tentacles are long and dark maroon. At the adult stage, the dark maroon colour of the tentacles starts to fade and the purple appears as stripes on the bell. The tentacles vary with the age of the individual, consisting typically of eight marginal long dark arms, and four central frilly oral arms. Often young Cancer crabs make home in the jellyfish and eat the parasitic amphipods that feed on and damage the jelly.

Cyanea capillata

Description: The lion's mane jellyfish (*Cyanea capillata*), also known as hair jelly, is the largest known species of jellyfish. These jellyfish are named for their showy, trailing tentacles reminiscent of a lion's mane. It may grow to over 2 m wide and its tentacles are up to 60 m long. These extremely sticky tentacles are grouped into eight clusters, each cluster containing over 100 tentacles, arranged in a series of rows. As in all jellyfish, the mouth of the lion's mane jellyfish is located on the underside of the bell-shaped body which is divided into eight lobes. It is often bioluminescent, emitting its own light. Size also dictates colouration – larger specimens are a vivid crimson to dark purple while smaller specimens are light orange or tan coloured.

Order: Rhizostomeae

Rhizostoma pulmo

Rhizostoma pulmo (Rhizostoma octopus) has a solid appearance. It varies in colour from whitish pale or yellow to shades of green, blue, pink or brown. Its umbrella margin is divided into a number of semi-circular lobes like extensions (marginal velar lappets).

It then divides into four pairs of oral arms that consist of three winged portions, followed by three winged elongated terminal appendages. Mature males possess blue gonads whereas, ripe females are reddish brown. When exposed to air, their nematocyst warts give the umbrella surface a matt like appearance. The custacean *Hyperia galba* may be found throughout the body of the jellyfish and more commonly in its gastric or gonad pouches. *Rhizostoma pulmo* has an average of ten velar lappets in an octant.

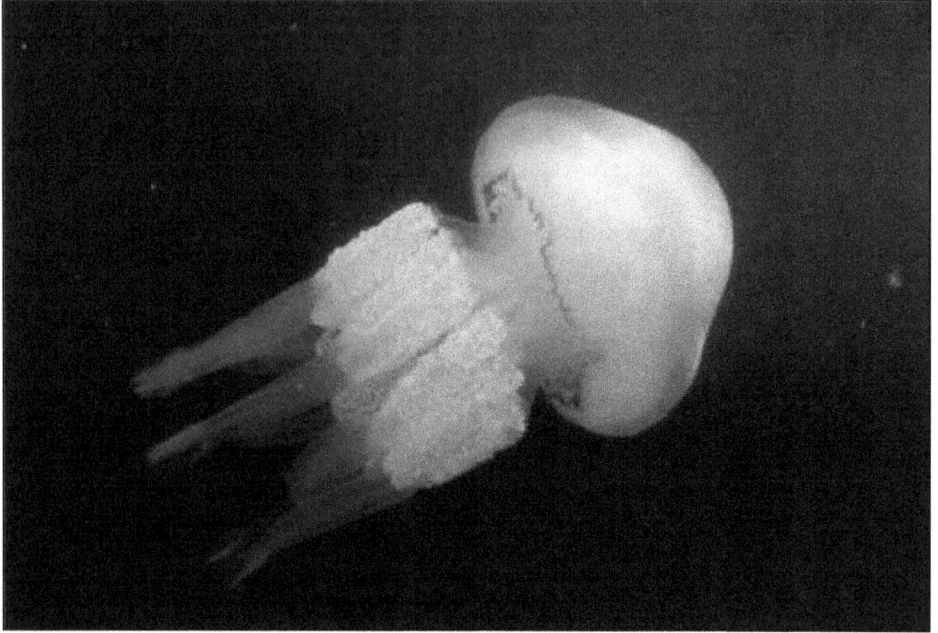

11.3. Acnidarians

Phylum: Acnidaria (Ctenophora)

Class: Tentaculata

Order: Cydippida

Pleurobrachia pileus

Description: This species has an egg- to spherical-shaped body; tentacular diameter slightly wider than sagittal; rows of ciliary combs equal in length, starting near aboral pole extending more than three quarters of distance toward the mouth. Tentacle sheats widely separated from stomodeum. Tentacles with tentilla simple.

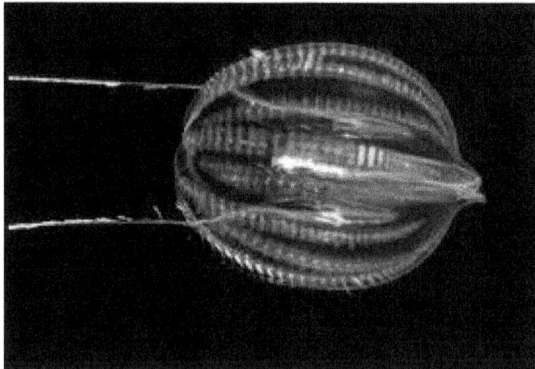

Colour: comb rows milky opaque, ectomesoderm glassy transparent, tentacle, sheaths and stomodeum milky or in some animals dull orange. Maximum length, 25 mm.

Class: Atentaculata (Nuda)

Order: Beroidea

Beroe sp.

Description: Conical ctenophores with very wide mouth and stomodeum; 4 meridional gasrovascular canals usually with numerous diverticulae. Without tentacles. Body mitten-shaped. Lateral compression very marked. Colour dull milky, pink or reddish- brown. Height up to 115 mm.

11.4. Marine Rotifers

The phylum Rotifera has about 2000 species of microscopic aquatic animals, with a length ranging between 80 and 1000 µm. Marine rotifers occur in both plankton and benthos, where they significantly contribute to the total biomass, and play a key role in the food chain process. In saltwater habitats, there are more than 500 rotifer taxa. A total of 37.0 per cent of marine rotifers have been found in saltwater only and are considered stenohaline. Some other taxa (39 per cent) are reported from both fresh and salt water habitats and are considered truly euryhaline. At least 24 per cent are known from freshwater habitats, and found in salt water only occasionally.

Phylum: Rotifera

Class: Monogononta

Order: Ploimida

Synchaeta sp.

Description: Body of the animal is sac-, bell- or cone-shaped with either lateral ear-like ciliar structures on the rotary organ or dorsally and ventrally two groups of long, leaf or blade-like flexible appendages. Feeding is mainly by grabbing-sucking. This species is found both in freshwater and marine systems.

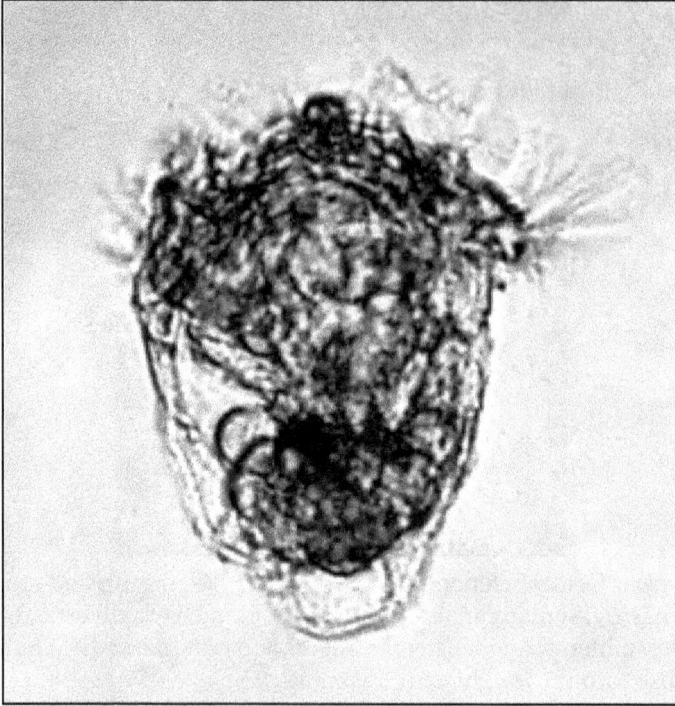

Brachinus calyciflorus

Description: Lorica saccate, rather flexible, flattened dorso-ventrally, smooth. Antero-dorsal lorica margin with 4 spines with broad base, pointed, posterior spines

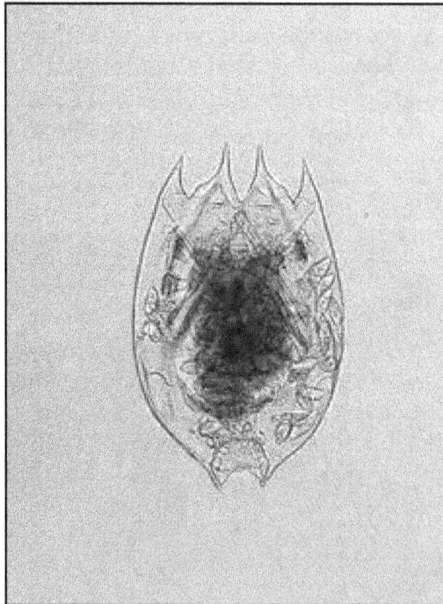

present or absent; with or without postero-median spines at foot opening. A very polymorphous species, with many variants. This species is preferred as test animals in aquatic toxicology because of its sensitivity to most toxicants. It is one of the live-food organisms used for the mass production of larval fish in aquaculture systems. It feeds on micro-algae and is a very euryhaline species found both in freshwater and brackish water systems and ubiquitous species. Lorica length 200600 µm.

Brachionus plicatilis

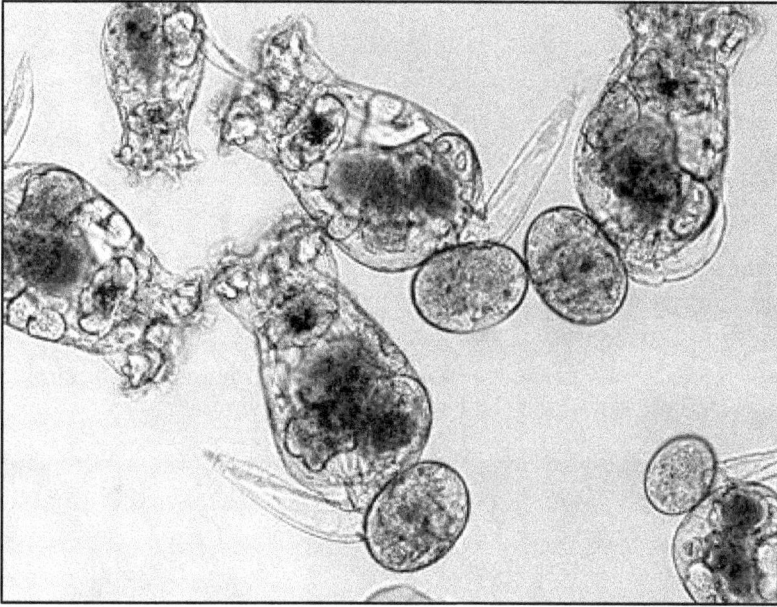

Description: Lorica more or less pyriform, soft, slightly compressed dorso-laterally, smooth or stippled. Antero-dorsal lorica margin with 6 short spines with broad base, nearly equal in length; median spines asymmetrical, with narrow anterior part, then rounding outward and forming broad base; intermediate spines more or less symmetrical or with outer half of base larger than inner one. Antero-ventral margin with 4 smoothly rounded lobes, or very broad lobes with shallow, blunt spines. Posterior spines absent. It is an euryhaline species occurring both in freshwater and brackishwater systems. Lorica length 275340 µm.

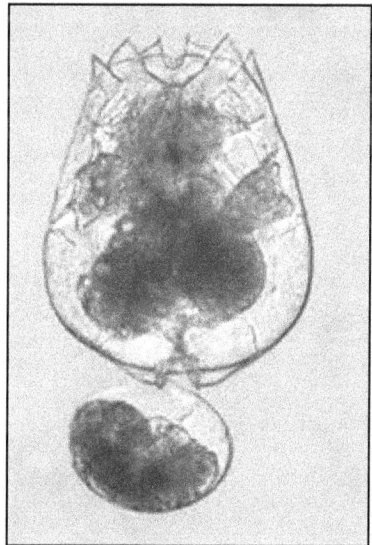

11.5. Planktonic Annelids

Pelagic polychaetes are a distinctive group of marine zooplankton, although they are less important in terms of species richness, abundance and biomass. Most species are epipelagic, inhabiting mainly the upper 50 m of the water column. Some, however, are mesopelagic or even bathypelagic. Pelagic polychaetes are widely distributed in all the seas and oceans of the world and are therefore considered as a cosmopolitan group.

Phylum: Annelida

Class: Polychaeta

Order: Phyllodocida

Plotohelmis capitata

Description: In this species, chromatophores are found on the ventro-lateral branching on the ventrum. Males with 4 pairs of genital papillae located on the ventral region of segments, 12 to 15.

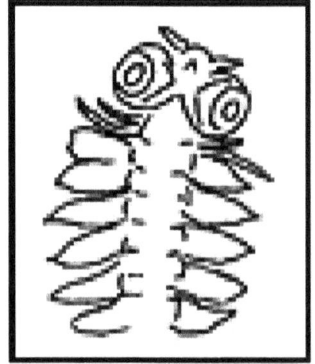

Pelagobia longicirrata

Description: Prostomium triangular with 4 small antennae and a pair of eyes. Tentacular segment with a few setae and 2 long, tapered cirri. Lateral parapodia with a conical setigerous lobe and longer dorsal and ventral cirri essentially similar, more or less cylindrical. Compound setae with basal joints smooth.

Tomopteris helgolandica

Description: Body relatively short and broad, up to 34 segments and a 'tail', which is absent in young individuals. Without obvious segmentation other than elongated parapodia. Prostomium with 2 short, divergent antennae, two eyes with lenses and two ciliated nuchal organs. The length of cirriform appendages of the second segment is at least half the body length.

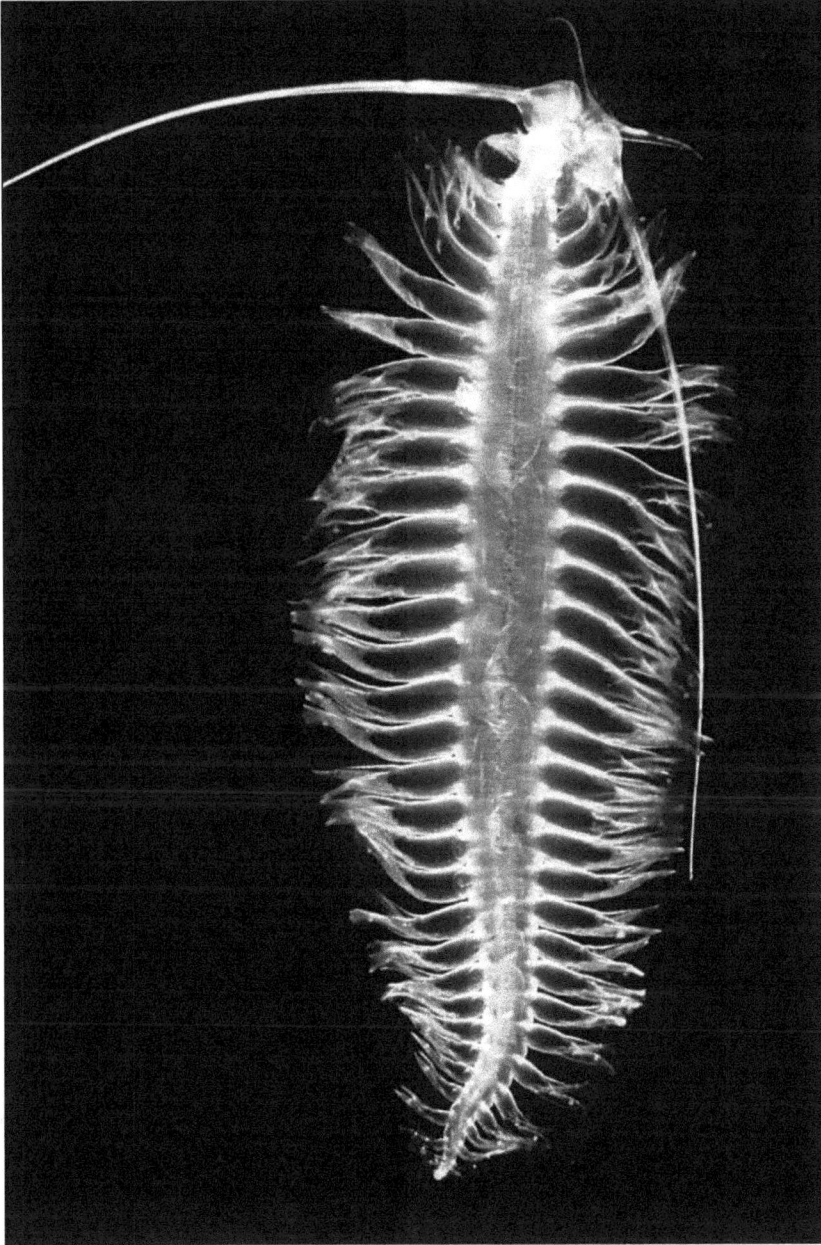

Anterior part of body with 14-20 parapodia and a 'tail' bearing 10-13 very reduced parapodia. Tail's length may be a third of the total body length. All parapodia with rosette organs in the tips of both branches. Parapodia without chaetae. Size of the animal 20-40mm in length, including the tail. Body striated with dark brown lines in life.

This pelagic, neritic species is capable of emitting yellow bioluminescence from its parapodia. Interestingly, this animal does not use luciferins for the purpose of emitting light.

11.6. Arrow Worms (Chaetognatha)

Arrow worms or glass worms are found both in deep hauls and surface coastal hauls. They are very transparent and long. The characteristic "hooks" of setae are found around the head. They are macroplanktonic and holoplanktonic carnivorous animals.

Sagitta enflata

Description: Number of hooks: 8-10; anterior teeth: 5-10; posterior teeth: 5-15. Maximum body length: 25 mm. Body flaccid, transparent; transversal musculature absent; head of medium width; hooks not serrated. Fin bridge absent. Anterior fins short, partially rayed, round; posterior fins short, partially rayed, round. Collarette absent; gut diverticula absent. Small eyes, with star-shaped pigment spot. Seminal vesicles round; position of seminal vesicles: touching tail fin, well-separated from posterior fins. Ovaries short, reaching to middle of posterior fins; ova large.

Sagitta maxima

Description: Body large and transparent. Two pairs of lateral fins, partially rayed; fins connected by a broad fin bridge. Anterior fins long, beginning the level of the ventral ganglion. Posterior fins rounded. Collarette and alimentary diverticula absent. Eyes small, with a small pigment spot.

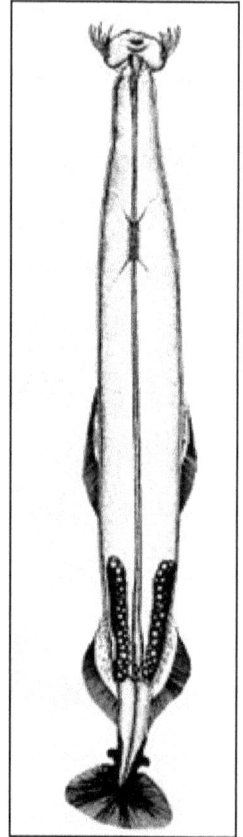
Sagitta enflata

Ovaries long, extending to the ventral region. Seminal vesicles round to oval in shape, located closer to the posterior fin. Hooks: 5-10. Teeth in two rows, anterior 4-5 and posterior 5-8. Maximum body length, 90 mm.

11.7. Cladocerans

The majority of water fleas are freshwater species with only a few species living in seas, but quite a large number particularly adapted for the brackish waters. Most cladocerans are small, less than 5 mm long. They have a carapace covering most of the body, except the head. No segmentation is visible on the carapace, but in many species, the carapace forms a spine posteriorly. Sometimes, there is also a spine on top of the head. The second antennae are very well developed and muscular and are used for swimming. Cladocerans have a large, sessile compound eye formed by the

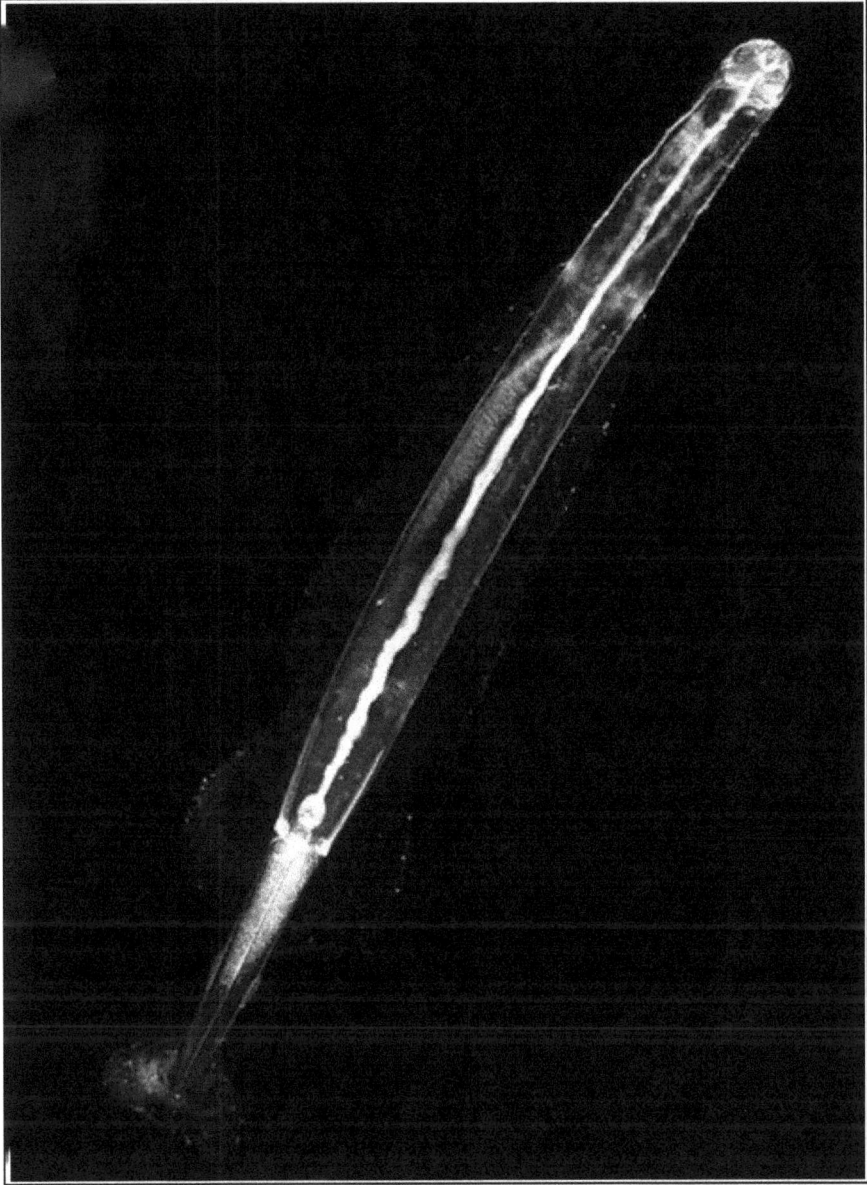

Sagitta maxima

fusion of paired eyes. The eye is capable of rotating in different directions. Typical cladocerans have 6 pairs of "trunk" appendages, and their bodies are not divisible into a separate thorax and abdomen. The tip of the trunk forms a "postabdomen", which is bent towards the ventral trunk surface and is equipped with claws and spines for cleaning the carapace.

Phylum: Arthropoda

Class: Crustacea

Order: Cladocera

Penilia avirostris

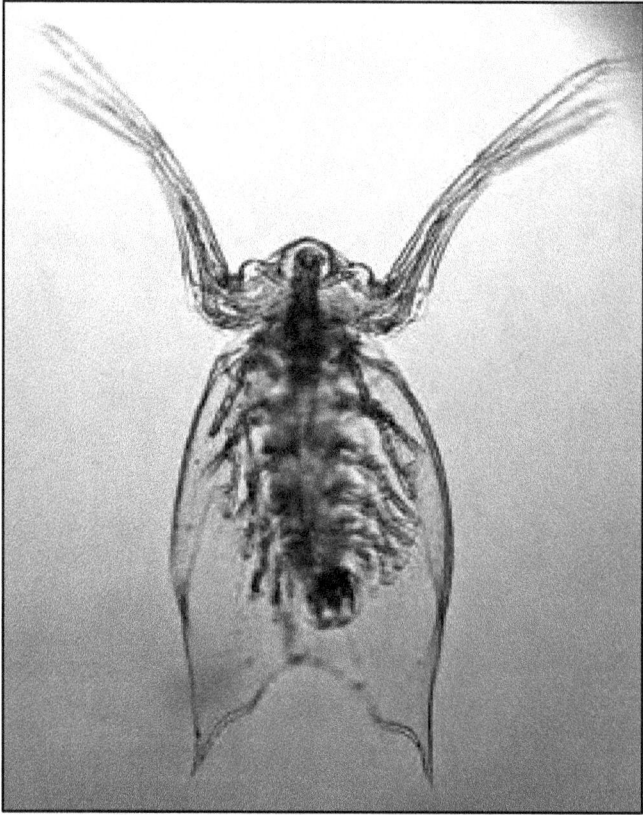

Description: This species is commonly seen in epipelagic zones. Free edges of carapace spined, the inferio-posterior angle of carapace has a larger spine. Six pairs of legs, the most posterior reduced. Antenna projected forward. Adult copulatory organs longer than the post abdomen. Female head with prominent rostral point; antennulae small and truncated. Male head rounded; antennulae as long as carapace. Male first leg with a strong distal hook. While female has a maximum size of 1.0mm, male 0.8mm. Coastal water species, rare in open oceans. Filter feeder on bacterioplankton and phytoplankton.

Podon polyphemoides

Description: Neritic, eurytherm, euryhaline surface-living species from eutrophic waters. Occurring at salinities ranging from 1 to 35ppt. Female with an almost spherical brood pouch, more or less perpendicular with respect to the body. Carapace

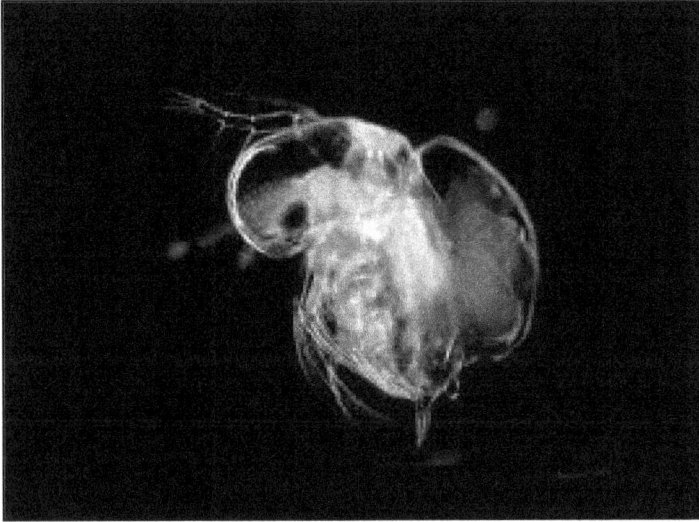

faintly reticulate. Head round. Eyes very large, occupying almost the whole forehead. Lower edge of labrum rounded. Antennulae are very small and are not able to move; antennulae bear at the end four sensory papillae. Postabdomen is very short with short end. Postabdominal setae are short but still much longer than the very small reduced caudal peduncle on which they are situated. Brood pouch with 2-6 embryos.

Male with small conical carapace. Head and eyes larger than in the female. The copulatory organ is rounded off distally. Colour grey to yellowish white, transparent; the postabdominal end is often bluish. Female 0.6 mm; male 0.5 mm.

Evadne spinifera

Description: It is a somewhat euryhaline species; Female egg-shaped in lateral view; brood pouch with a large, marked spine. Antennulae clearly separated from the lower side of the head, lower distal side almost perpendicularly curved upwards. Legs rather short, the first leg not elongated. Leg 4 very short, not well segmented, with four small distal bristles; Caudal peduncle well developed. Brood pouch with 6-7 parthogenetic eggs.

Male carapace narrowing and pointed strongly in lateral view, however, without marked spine. Colour of the species ranges from bright grey to bright yellow, opaque. Female 0.9-1.30 mm; male 1.0 mm.

11.8. Ostracods

Ostracods are small crustaceans which occur in most aquatic habitats. They range in size from 0.5mm to 30mm. Normally marine planktonic ostracods are rare in the inshore waters, and they occur almost everywhere in the deeper regions. They are usually the second-most abundant group of macroplanktonic animals and second only to the copepods. Some of the dominant species of planktonic ostracods are considered to be the most abundant invertebrates in the world.

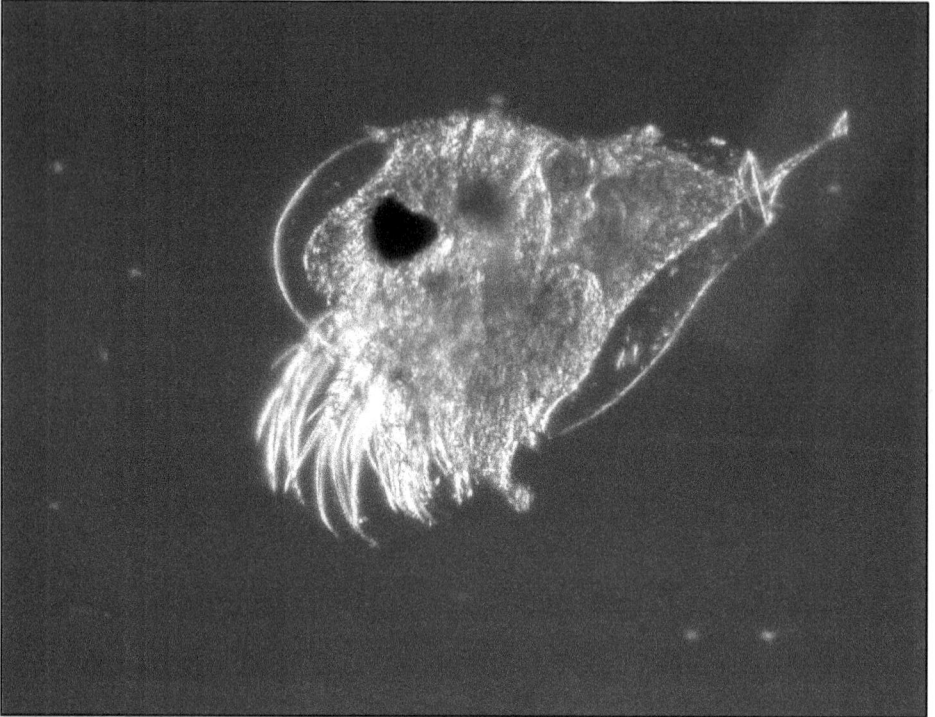

Evadne spinifera

Class: Ostracoda

Order: Myodocopida

Cypridina meditteranea

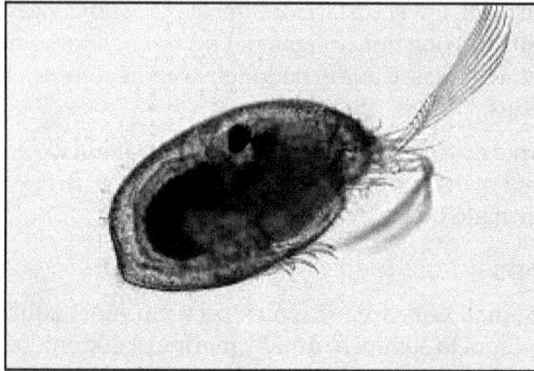

Description: This animal is a nocturnal visitor of the plankton - coming normally from benthos. Body is 1. 6 mm long. There are 2 antennae. In the middle of the animal is the eye.

Cypridina acuminata

Description: Mature females measure 2.0 mm and the height of the shell is about 50 per cent of the length. Mature males measure 2.3 mm. The rostrum is rounded antero-dorsally and pointed inferiorly, the tip may carry 1-2 minute teeth. Medially across the rostrum is a row of 4-5 bristles and close to its tip 4 more. At the bottom of the incisur are one larger and one shorter bristle. On the shell margin just below the incisur are 4 long hairs.

Cypridina dentata

Description: Length of carapace in this species varies from 1.5 to 2.0 mm. Shell is distinguished by prominent, rounded, anterior corner and the dorsal anterior corner of rostrum is very conspicuous, acute and the posterior process is broader and shorter. The serrature of the anterior margin below the incisur is due to a number of plates along the edge of the shell and these plates are rectangular with rounded angles. Medially on rostrum are 4-5 bristles. The ridge in front of the posterior process is almost straight, it has a row of 15 spines on the right shell and 12 on the left.

Cypridina sp.

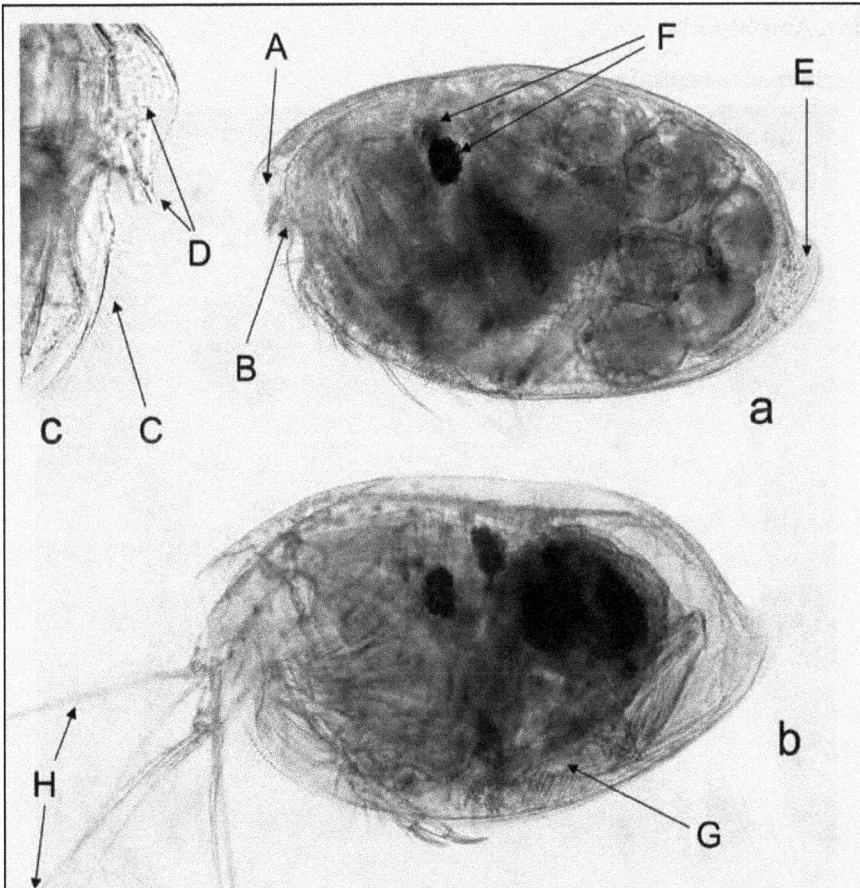

Description: Carapace is rather elongate, height is about 60 per cent of total length. Anterior part of rostrum is protruding over the front margin; Carapace margin below the incisures (B) is either bare or with a row of 2 to 25 hairs (C); the number of these hairs and also the spines at the rostrum and incisure edge (D) are good distinguishing characters. Posterior carapace process (E) always well developed. Labrum has anterior part with two unpaired processes and posterior one with two paired processes. Each furcal plate arms by 10 claws. The males differ easily from the females by the presence of two very long setae on first antenna (H,). These setae are about twice the carapace length. Length: Females 1.35 mm, males 1.35-1.40 mm.

11.9. Amphipods

Amphipods are abundant and widespread crustaceans, found in all marine and freshwater habitats. They are beneficial to ecosystems in marine food chains, as herbivores, detritivores, and scavengers. Amphipods are poor swimmers and they lack a caprapace, have sessile eyes, the pleopods are respiratory in function, and the uropods do not form a tail fin. These animals are laterally flattened and there is a division of functions among the varied limbs. Extraordinarily abundant in the rocky coastal regions of all seas.

Order: Amphipoda

Hyperia macrocephala

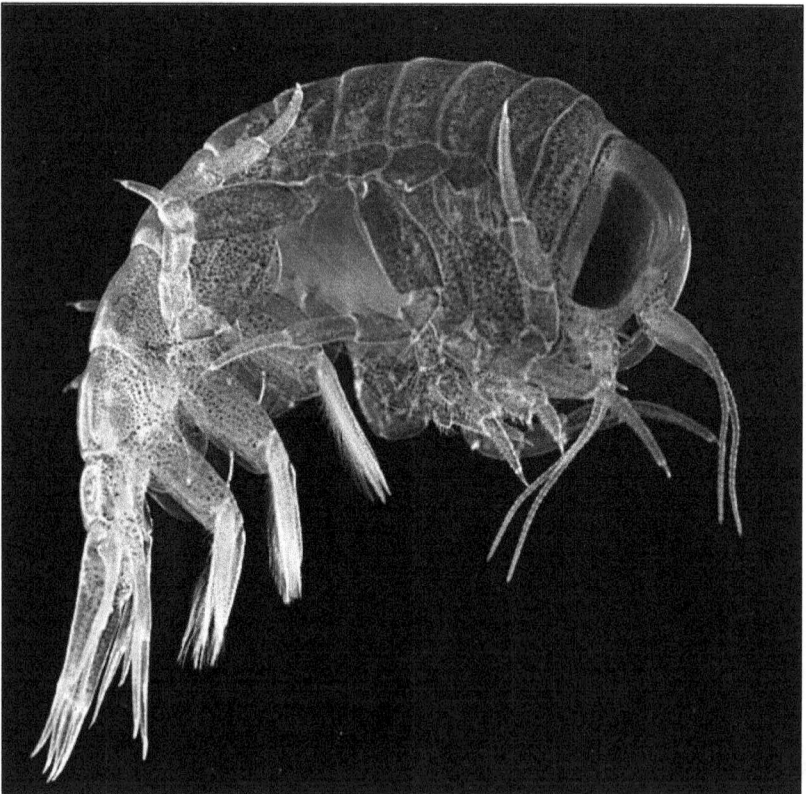

Description: Head of this species is subequal to first two pereonites combined. Epimeral plates are with sharply pointed posterodistal corners. Telson is moderately short. Length, 28 mm.

Hyperia galba

Description: Body plump, rounded; maximum length 12 mm; light translucent brown, with enormous green eyes. Head rounded, short and deep, with eyes occupying the whole of the two sides. Pereion is deep, rounded laterally and dorsally and somewhat compressed laterally; Pleon is slender and the urosome well developed. Female with very short, subequal antennae. Male with long slender antennae; antenna 2 longer than antenna 1, about two-thirds of the body length. The maxilliped lacks a palp.

Gnathopods 1 and 2 small, simple; carpus with an acute, projecting, disto-ventral lobe; pereiopods 3 to 6 are typically slender and in all pereiopods, the coxa is very small. Pleopods are well developed. Uropods 1 to 3 broad, laminar; rami lanceolate, with finely denticulate margins.

11.10. Copepods

The copepods often dominate the plankton community. They are generally small in size. They form a link in the food web between the primary-producing phytoplankton and the plankton-feeding fish. Most of the economically important fish depend on copepods and other zooplankters during their larval stages as well as adult stages of some fish such as *Pampus argenteus, Tenualosa illisha,* and *Liza subviridis.* There are about 2,000 known marine pelagic species of copepods. The copepod females lay eggs freely into the water, or produce external paired or single egg sacs. The eggs hatch into copepod larvae. The first larvae are called nauplii. Larval copepods usually pass five or six naupliar stages, which are separated by a moult. The 6th naupliar stage moults into the first copepodite stage, which largely resembles the adult copepod. After molting through 5 copepodite stages, copepods attain adulthood and cease moulting. A typical copepod may undergo ecdysis (moulting) twelve times in its life.

Copepod Feeding

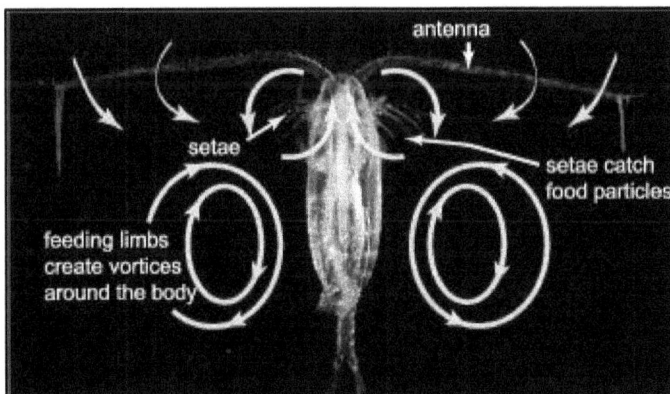

Feeding Mechanism in Copepod

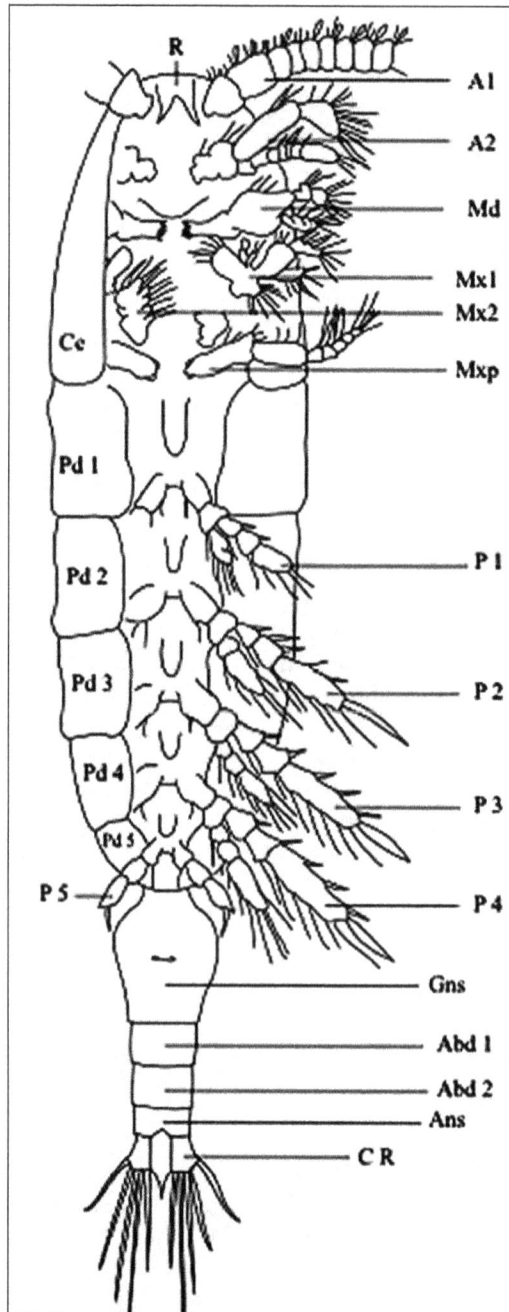

Parts of an Adult Calanoid Copepod

A1: Antennule; A2: Antenna; Abd 1: Abdominal somite 1; Abd 2: Abdominal somite 2; Ans: Anal somite; CR: Caudal ramus; Gns: Genital somite; Md: Mandible; Mx1: Maxillule; Mx2: Maxilla; Mxp: Maxilliped (with coxa, basis and endopod); P1-5: Swimming legs 1-5; and Pd 1-5: Pedigerous somites-5.

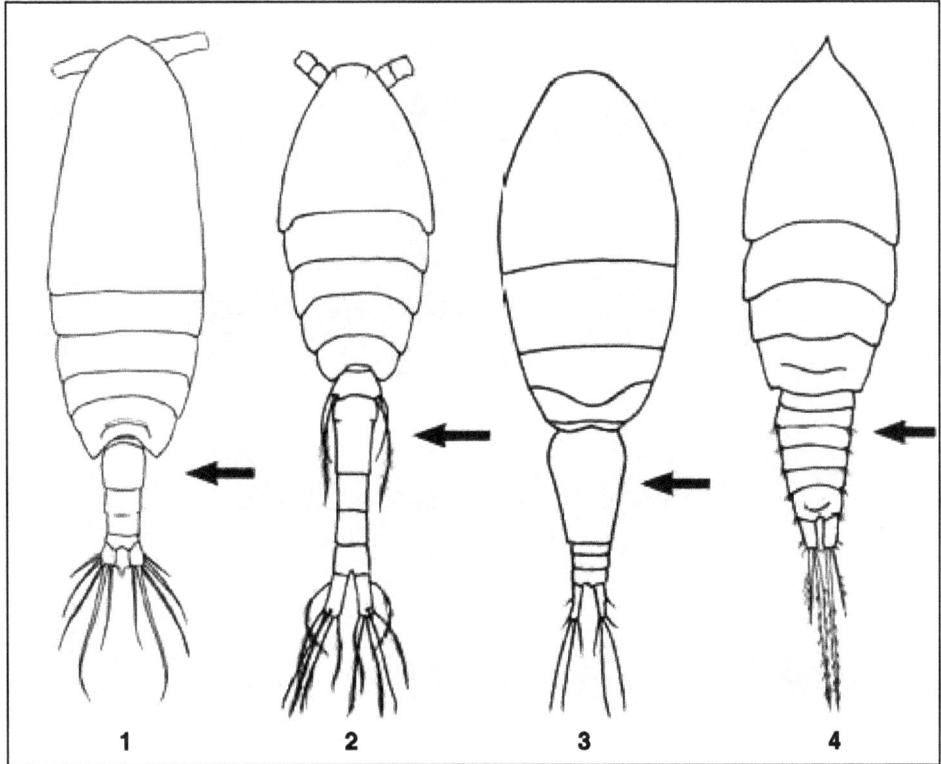

Morphological Variations in Copepod Suborders.
1: Calanoid; 2, 3: Cyclopoid; 4: Harpacticoid.

Copepods generally feed on phytoplankton. The beating of the feeding appendages pushes water postero-laterally, forming a large swirl on each side of the animal. Some of this swirled water is sucked antero-medially by the outward swing of the maxillipeds. The inward swing of the maxillipeds then pushes water between the setae and setules (bristles on the setae) of the second maxillae, which sieve particles out of the water. The filtered water is then expelled anteriorly by the first maxillae, and the captured food is taken to the mouth by the endites of the first maxillae.

Order: Copepoda

Sub-Orders of Copepoda

Calanoida: The species of this group have a major movable articulation between the 5th and 6th thoracic segment, and posterior to their 5th thoracic legs. They are more oval shaped and often have long or branched antennae.

Cyclopoida: The species of this group have a major movable articulation between the 4th and 5th thoracic segment, or between their 4th and 5th thoracic legs. They are more round shaped and often have shorter or less branched antennae.

Harpacticoida: The species of this group have a metasome that is more or less the same width as their urosome. They often have very long setae extending from their caudal rami.

Suborder: Calanoida

Acrocalanus longicornis

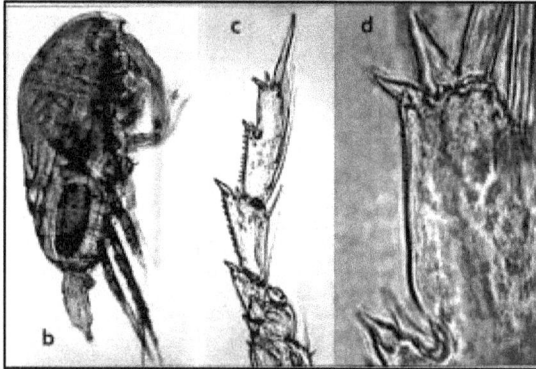

Female: Dorsal surface of the body moderately strongly arched, anterior head rounded. Length 1.00 – 1.26 mm

Male: Length 0.95 – 1.25 mm.

Paracalanus indicus

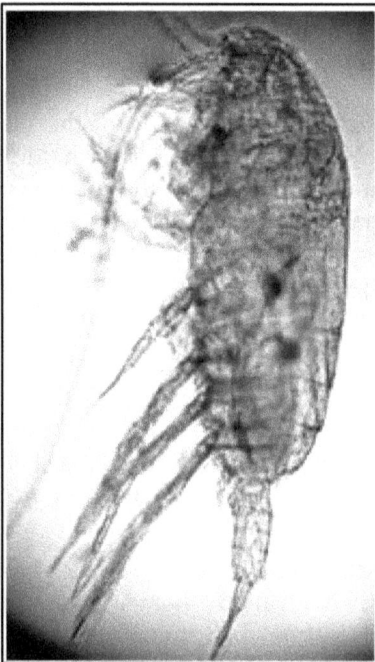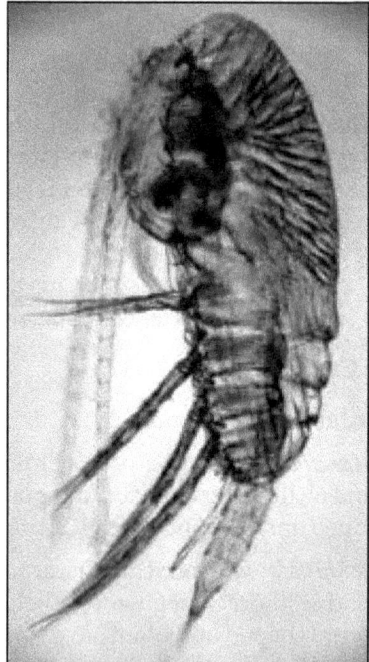

Female   **Male**

Female: Body stout with wide, smoothly rounded head. Prosome 2.5 times longer than wide. Rostral filaments relatively long. Length: 0.8-1 mm.

Male: Urosome is 5 segmented. Length: 1 mm.

Subeucalanus flemingeri

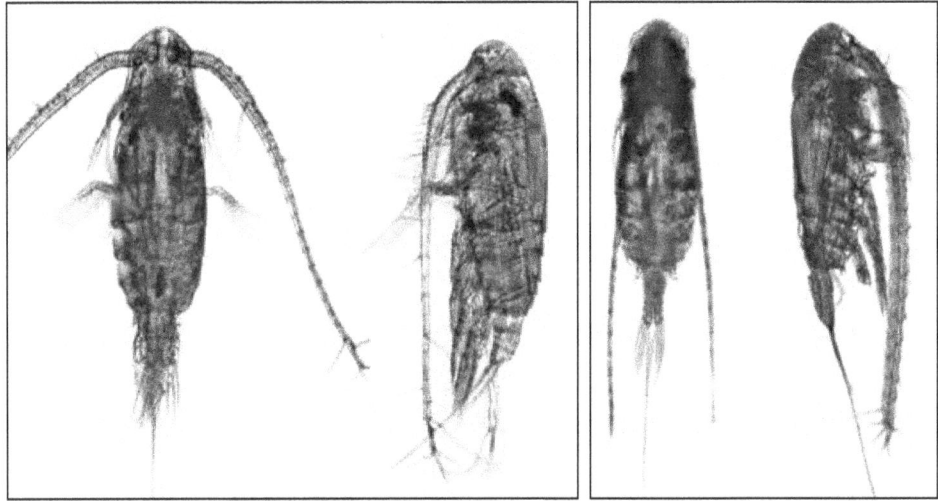

Female   **Male**

Female: Length 2.01-2.14 mm. Prosome nearly 3.1 times as long as wide and 5.7 times urosome. Forehead rounded. Rostrum robust, elongated, and ending in fine filaments slightly swollen at their bases. Genital somite broader than long with widest part in dorsal view on the posterior half of the somite.

Male: Length 1.97-2.14 mm. Prosome nearly 3.1 times as long as wide and 4.1 times as urosome. Forehead rounded. Genital somite slightly asymmetrical, with genital aperture located on left side; second somite the largest and cask-shaped.

Euchaeta rimana

Female: Length 2.80-4.30 mm. Posterior thoracic corners slightly asymmetrical, more produced on right. Left side of genital somite in dorsal view without prominent projection, right side with moderate projection that obscures more ventral projection. Genital field in ventral view with genital pads almost meeting in midline, right pad triangular.

Male: Length 3.11-4.10 mm. Thick spine arising at base of serrate lamella less than half length of lamella.

Centropages furcatus

Female: Length 1.68-1.78 mm. Head with a ventral ball-like eye. Posterior borders of prosome pointed and have an accessory spine on the interior border of this spine. Genital somite without spines on ventral surface; analsegment is not quite symmetrical.

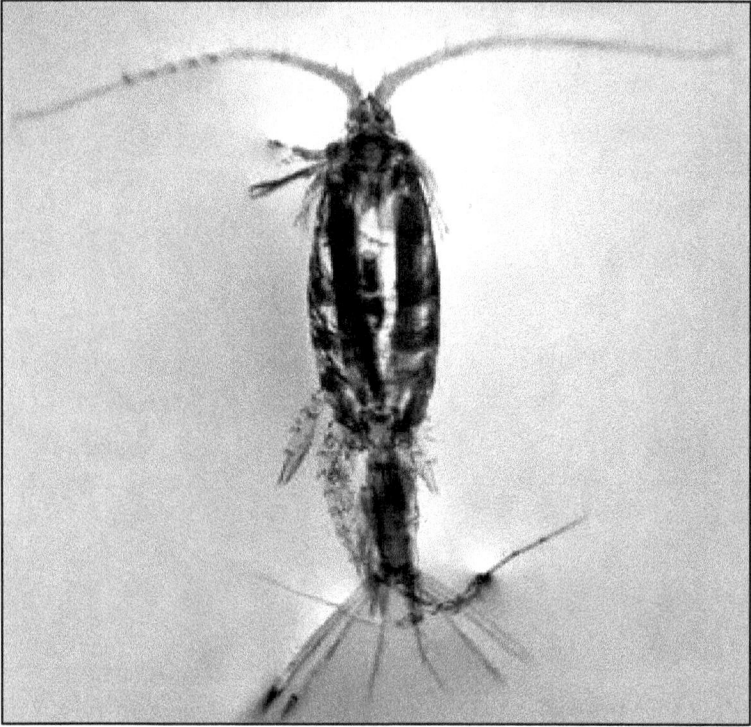

Euchaeta rimana

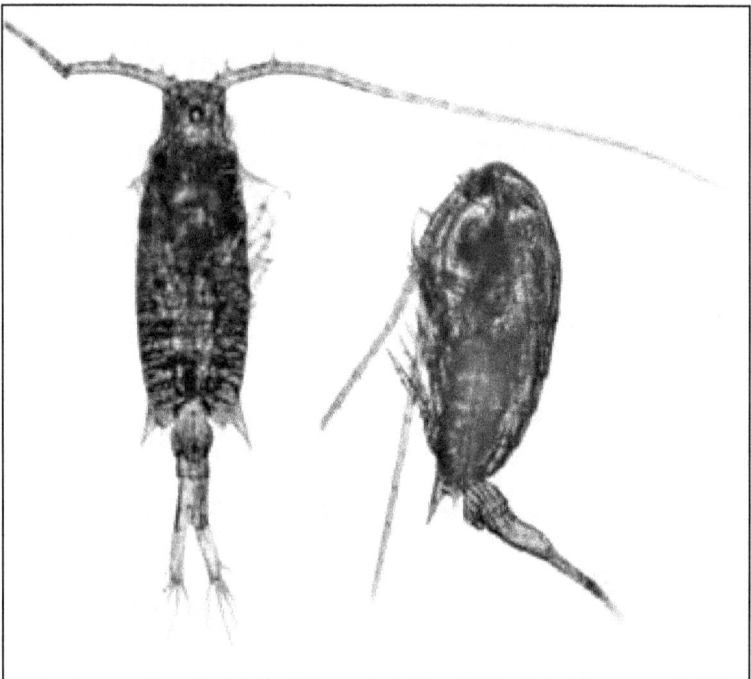

Centropages furcatus–Female

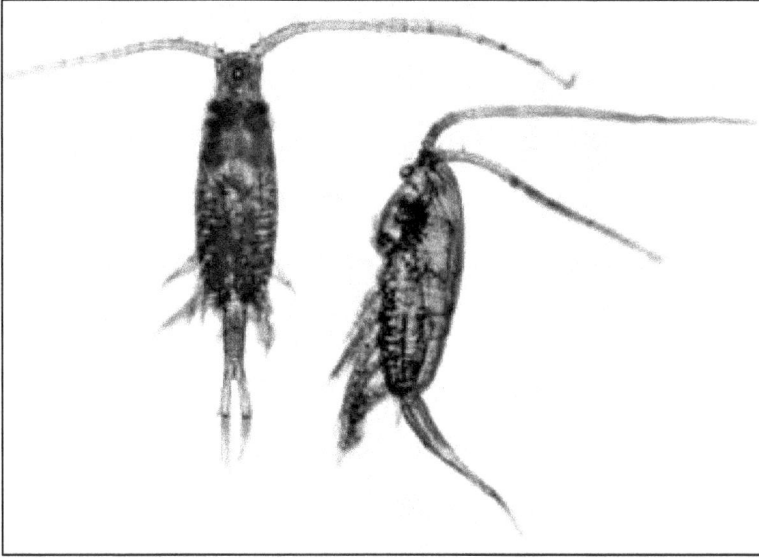

Centropages furcatus–Male

Male: Length 1.54-1.66 mm. Prosome posterior border slightly asymmetrical, pointed, with an accessory spine on the interior border; left side more protruding than right. Caudal rami slender.

Centropages indicus

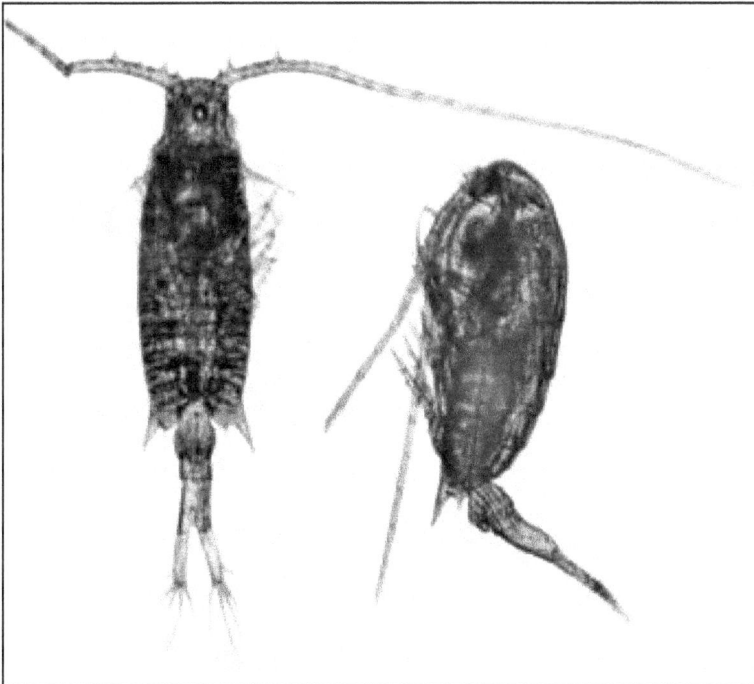

Female

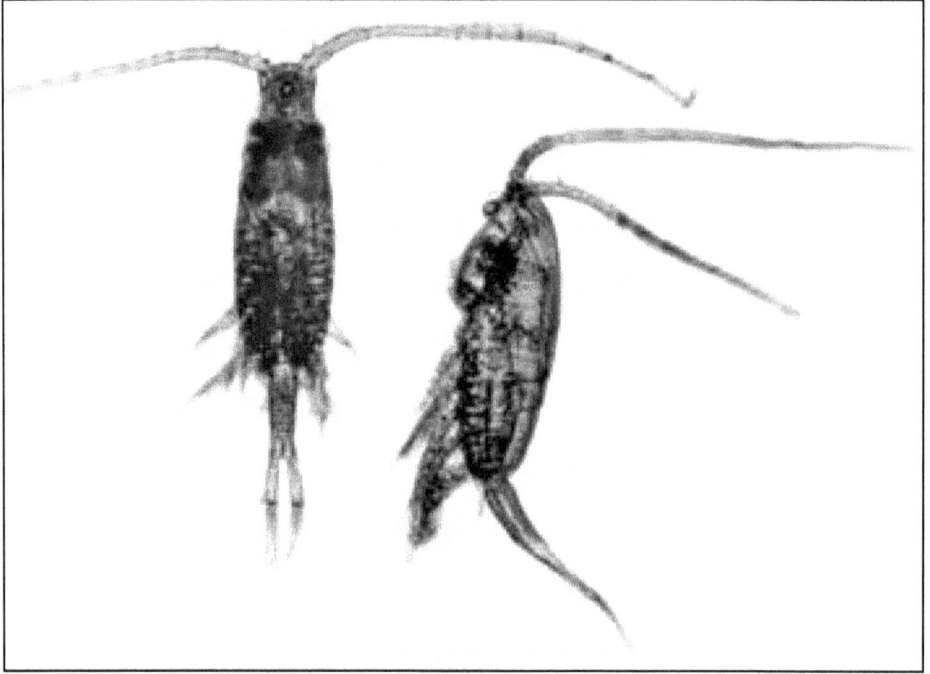

Male

Pseudodiaptomus arabicus

Female: Urosome 4-segmented. Urosome segment 1 with paired genital flaps, egg sac single. Leg 5 symmetrical, Length: 1.2-1.7 mm.

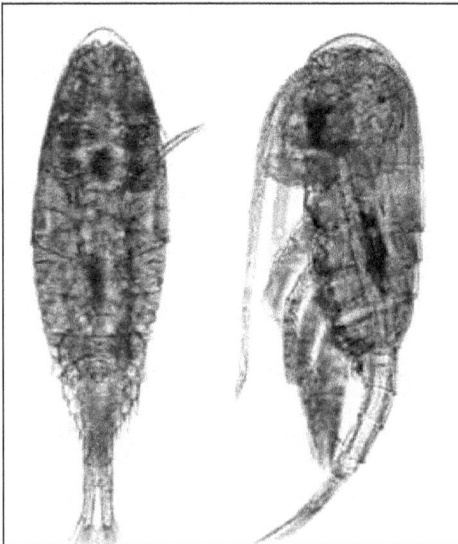

Female   **Male**

Male: Urosome of 5 segments;. Leg 5 asymmetrical,; left leg 2-segmented; both legs with endopods. Length: 1.1-1.45 mm.

Temora discaudata

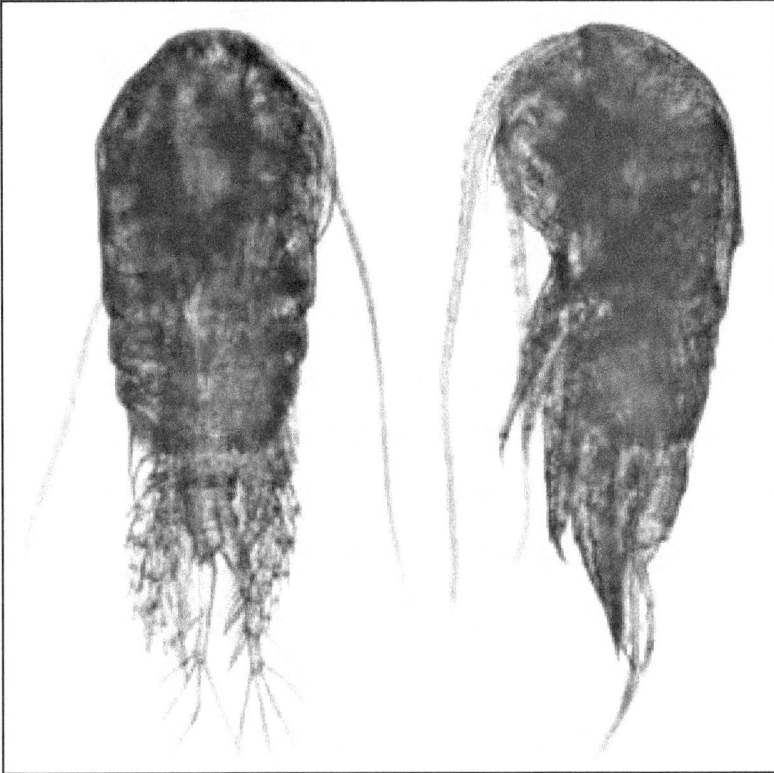

Female

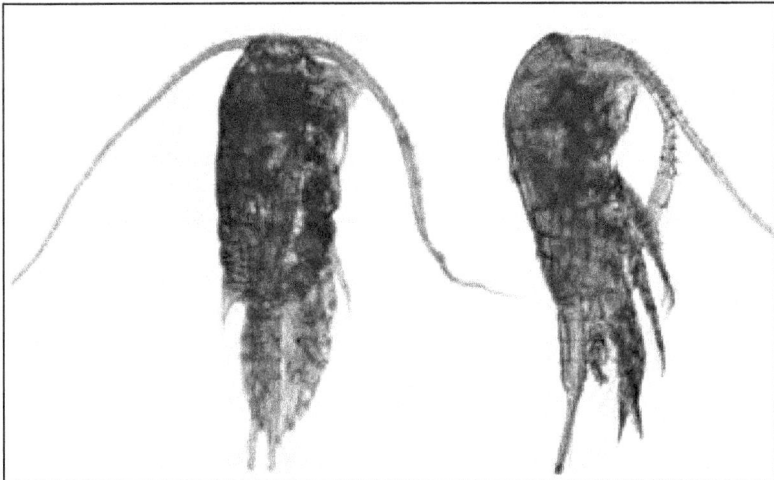

Male

Female: Quite a large robust species; posterior metasome segments produced into spines. Posterior corners of prosome pointed; Ur3 and caudal rami hardly asymmetrical right caudal rami slightly bent and thickened. Long slender caudal rami asymmetrical. Length: 1.6-2.0 mm.

Male: Lateral angles of the last metasome segment pointed and slightly asymmetrical;. Urosome of 5 segments, almost symmetrical, with long slender caudal rami. Length: 1.7-1.85 mm.

Temora turbinata

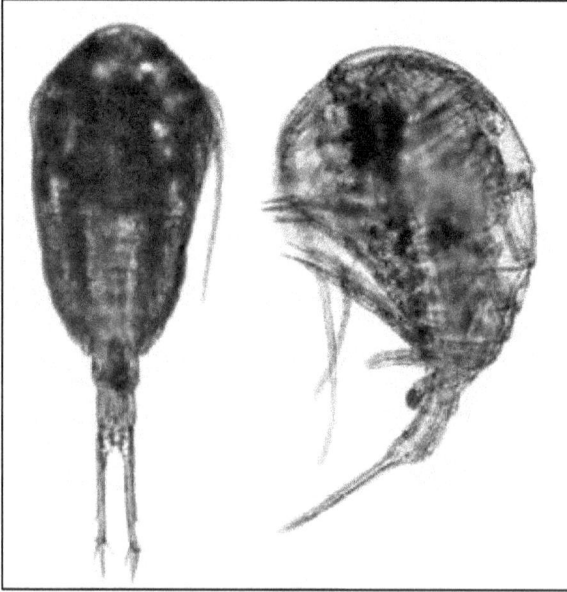

Female

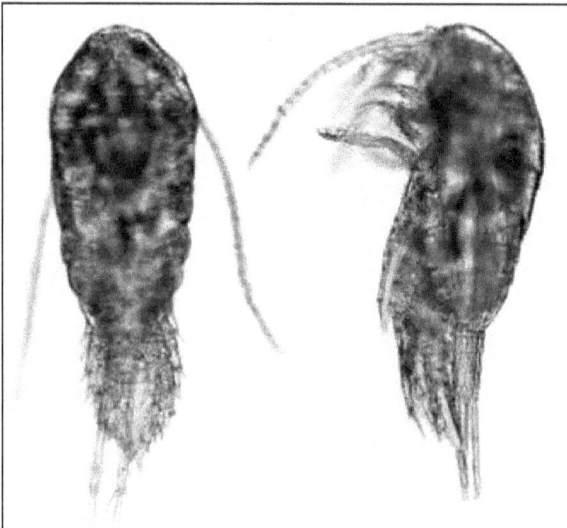

Male

Female: Body widest at cephalosome, tapering anteriorly to urosome; last metasome segment rounded.Urosome 3-segmented; anal segment symmetrical, shorter in length than previous segment; caudal rami almost symmetrical, very long and slender. Length: 0.9-1.6 mm.

Male: Body shape similar to **Female:** Urosome quite symmetrical; caudal rami long and slender; Length: 0.9-1.4 mm.

Calanopia elliptica

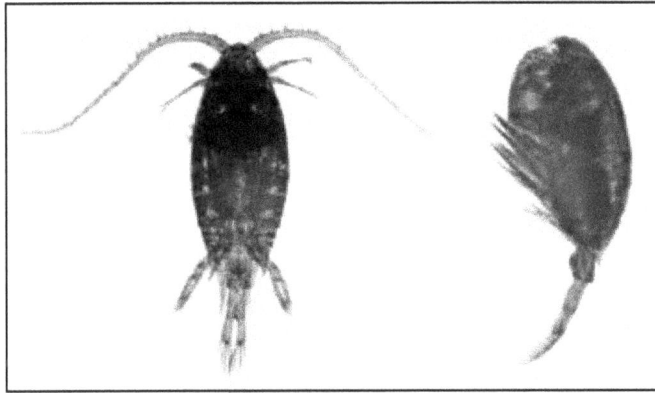
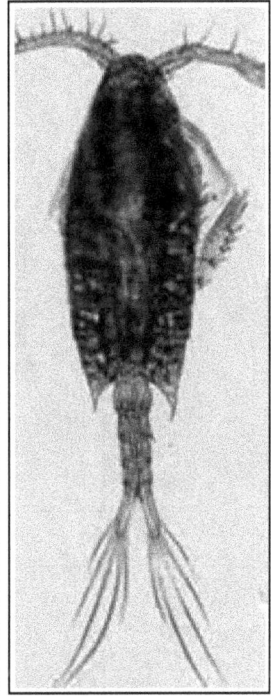

Female **Male**

Female: Cephalosome enlarged, conically rounded. Rostrum bifid, slightly bulged at ending in acute points. The prosome nearly twice as long as the combined length of the urosome. Long points on posterior metasome. Caudal rami nearly 3 times as long as broad. Length: 1.6-1.9 mm.

Male: Urosome segment 2 right side distal border produced into a well defined tooth. Length: 1.4-1.8 mm.

Labidocera acuta

Female: Length ratio prosome to urosome 2.94:1. Body elongated ending in a single spiniform process, cephalon rounded with a median anterior hook, dorsal eye lenses moderate, separated. Lateral cephalic hook absent. Length 2.88 mm.

Male: Length 2.65 mm, length ratio prosome to urosome 2.96:1. Body similar to female, anterior hook more pronounced, dorsal lenses slightly larger and closer together than in female.

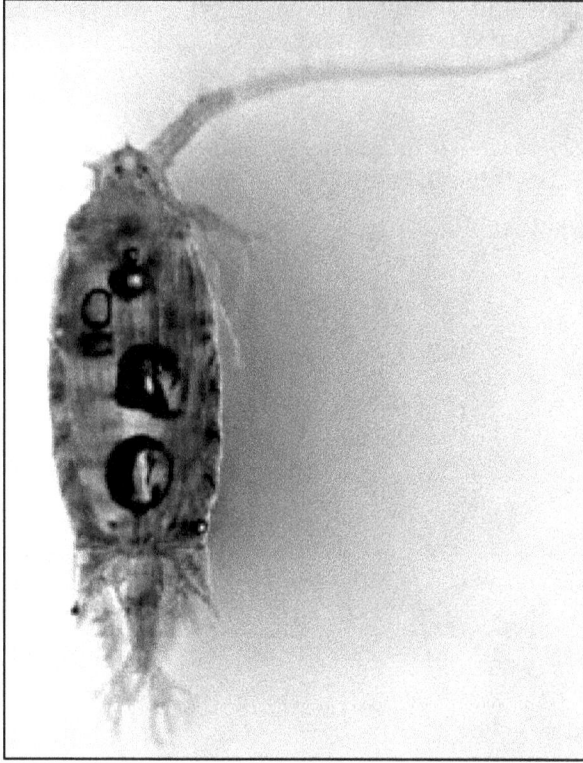

Female

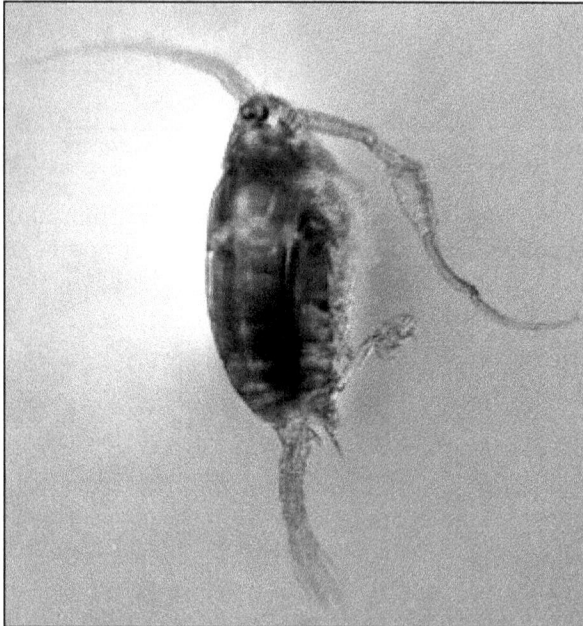

Male

Pontella danae var. ceylonica

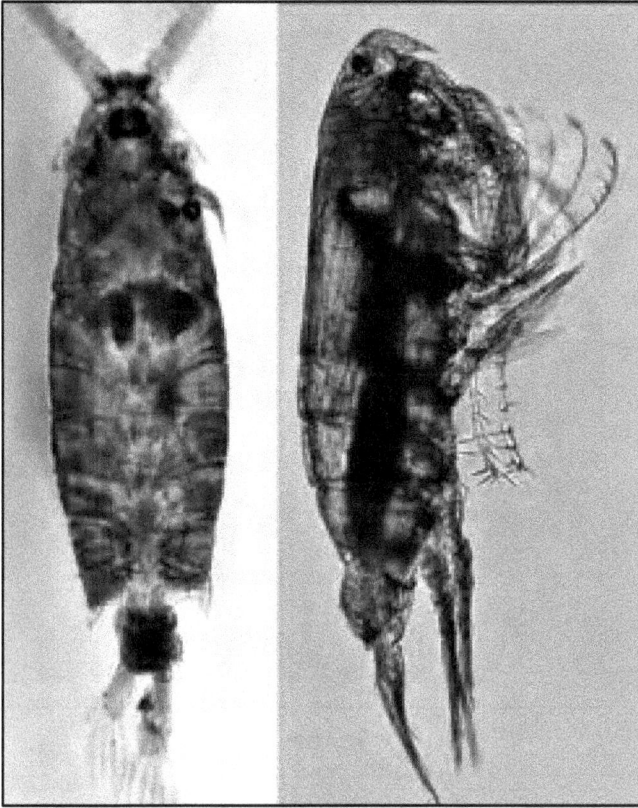

Female

Female: Length: 2.7-3.2 mm. Cephalosome with dorsal eye lenses and ventral lenses well developed; rostrum bifid, tapers to tip; rostral lens feebly developed; prosome more or less of same width throughout. Caudal rami asymmetrical, right ramus distinctly broader and longer and carries a marginal fold mediodorsally, middle three setae of right ramus bulbous at bases.

Pontellopsis herdmani

Female: Length: 1.7-2.1 mm (1.9-2.32 mm). Body stout, rostral base prominent on dorsal view. Urosome asymmetrical and of two segments. Genital somite longer than segment 2 of urosome; segment dorsally towards mid-dorsal margin on the left side with a small spine and with a well developed spine on its right distal corner; a short indentation present on its left distal corner; a short indentation present on its left distal corner. Caudal rami slightly asymmetrical, right one being perceptibly broader than left. P5 symmetrical; exopod with two small outer marginal spines and with subequally bifid tip; endopod bifid at its tip, both rami curved inwards.

Female

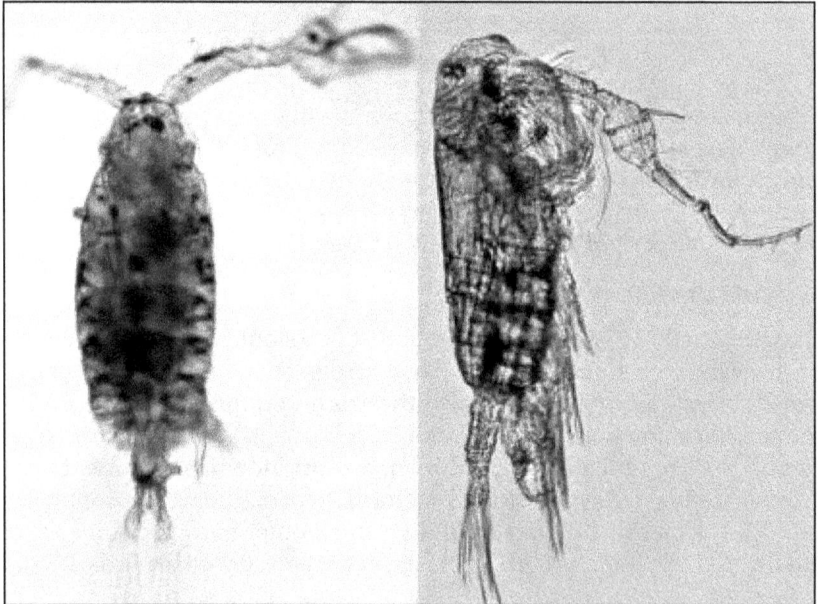

Male

Male: Length: 1.5-1.9 mm Cephalosome resembles that of femal.e Posterior corners of metasome 5 modified into a rounded lobe on its left side and into an acuminate long spine on its right side, which is curved inwards; Right antennules geniculate; thumb of chela is long, exceeds length of claw and is serrated at its apex; left leg with terminal segment carrying two subequal spines distally and one outer marginal spine;inner margin with hand of setae.

Acartia (Acanthacartia) fossae

Female: Length: 0.9-1.2 mm Posterior metasome segment with a row of 4-5 tiny spinules on the dorsoposterior margin.

Male: Length: 0.9-1.1

Acartia erythraea

Female: Posterior margin of the metasome is drawn out into prominent spines. Two smaller spines are also situated dorsally near the middle line. Urosome is 3segmented.Second urosomal segment bears small spines. Length1-1.5mm.

Male: Urosome is 5 segmented. Second urosomal segment is wider than long and bears 2 pairs of prominent spines.Length 1-1.3mm.

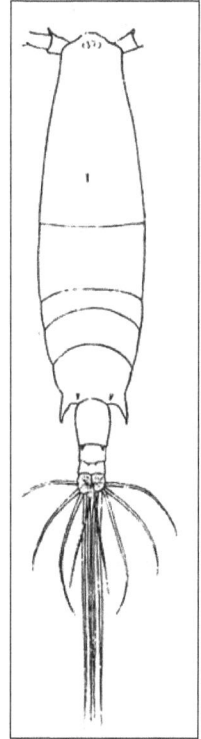

Acartia erythraea

Tortanus discaudata

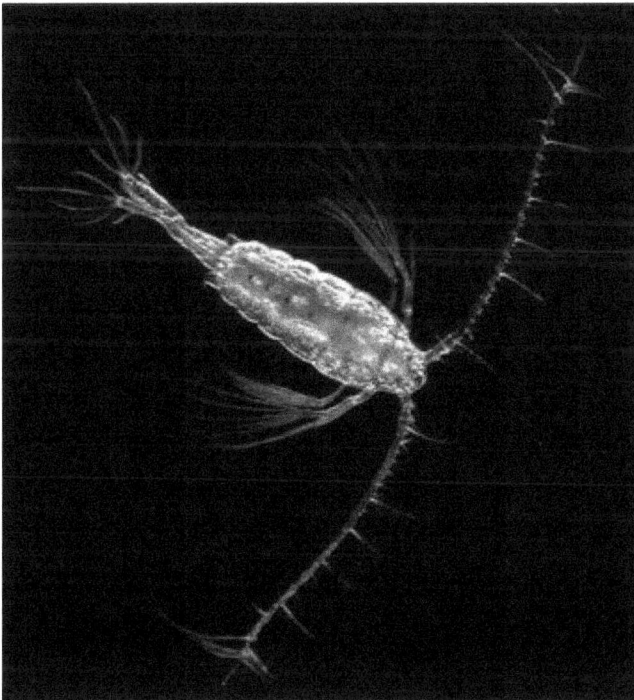

Female: Length: 1.15-1.5 mm Urosome of 3 somites; anal somite and caudal rami fused, asymmetrical;

Male: Length: 0.9-1.2 mm.

Tortanus barbatus

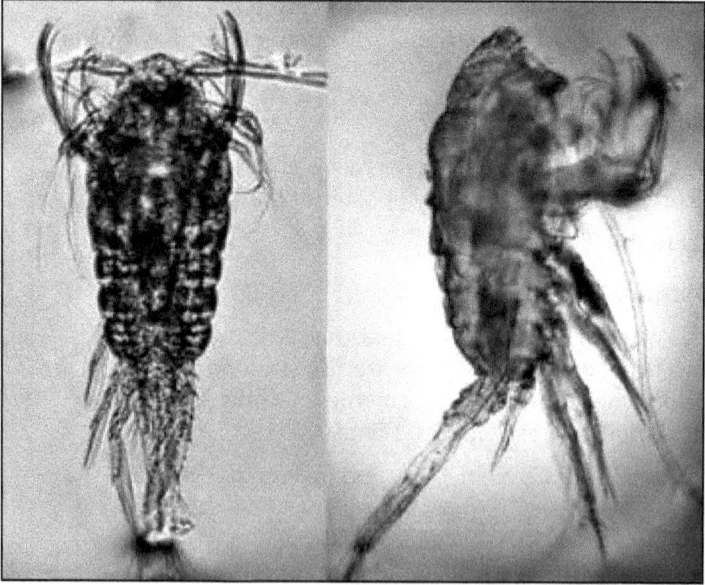

Female

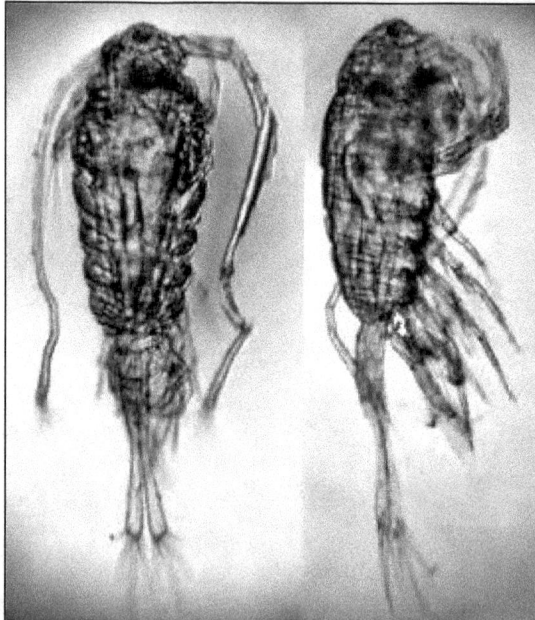

Male

Female: Length: 1.15-1.5 mm (1.32-1.6 mm). Urosome of 3 somites; anal somite and caudal rami fused, asymmetrical;

Male: Length: 0.9-1.2 mm (1.02-1.12 mm). Anal somite and caudal rami fused, slightly asymmetrical. terminal segment with a long outer seta, with numerous spinules on the inner and outer distal edges.

Rhincalanus cornutus

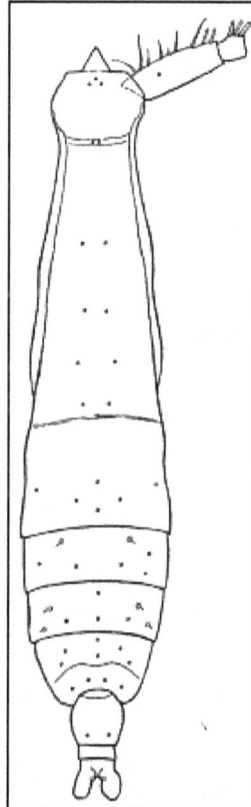

Male: Fifth leg is uniramose on the right and biramose on the left. Anterior projection of the head is distinctly anchor-shaped and the rostral filaments are visible in dorsal view. Length, 2.7mm;

Female: Resembles male in general morphology except in the 5th pair of legs. Length, 3.6mm.

Eucalanus attenuatus

Female: Head end is bluntly pointed and indented on either side of the frontal margin. Uroosme is 3 or 4 segmented. Caudal rami are fused to the anal segment. Fifth legs are absent. Length 4-6 mm.

Male: Head is triangular but less pointed than in the female. Length 3-3.5mm.

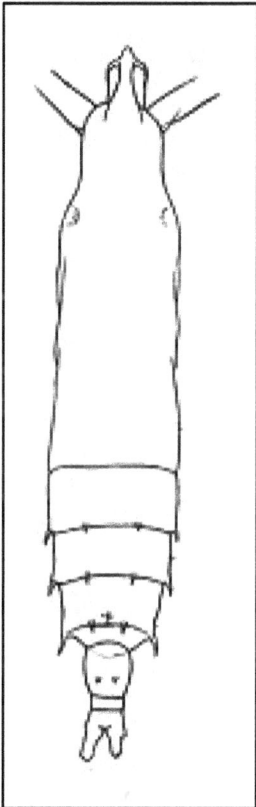

Rhincalanus cornutus   **Eucalanus attenuatus**

Suborder: Cyclopoida

Oithona attenuata

Female: Length: 0.6-0.8 mm. Prosome rounded rhomboid, greatest width at posterior end of cephalosome; length 1.5 times width, 1.0 times urosome. Head in lateral view bent ventrally into small process; lateral seta not extending to posterior margin of caudal rami.

Male: Length: 0.5-0.6 mm Prosome ovoid; length 1.9 times width, 1.2 times urosome. Head in lateral view bent ventrally into small process.

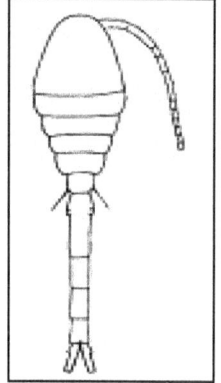

Oithona plumifera

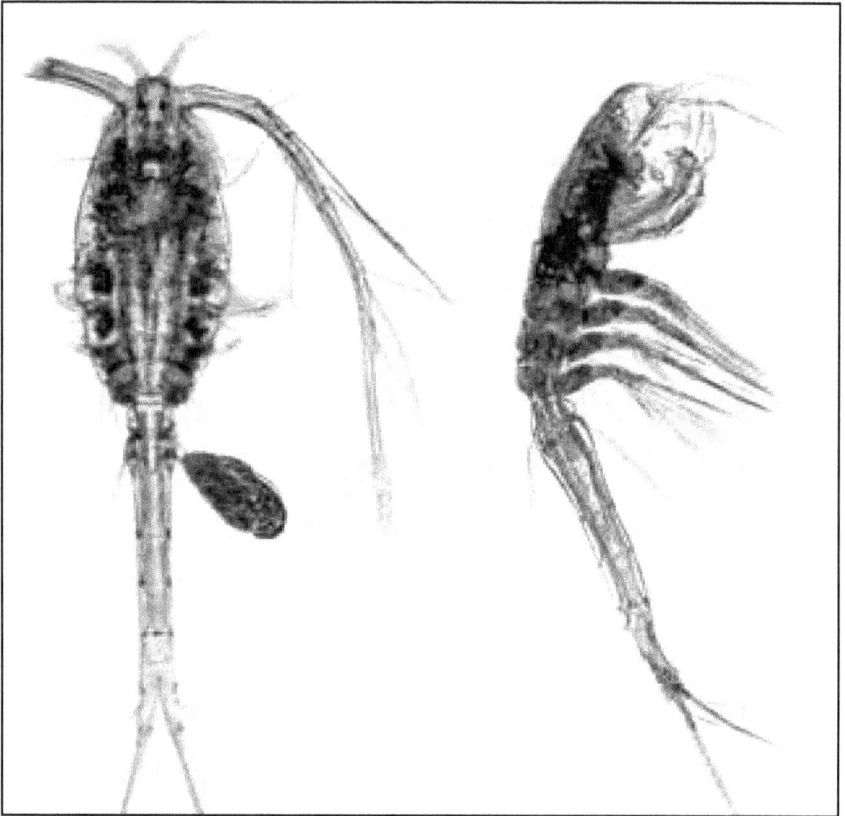

Female

Female: Length: 1.0-1.3 mm Prosome fusiform in dorsal view; length 2.4 times width, 1.05 times urosome. Head narrowing anteriorly in dorsal view; in lateral view bent anteroventrally into sharply pointed rostrum. Caudal rami lateral seta extending beyond posterior margin of caudal rami.

Male: Length: 0.75-0.8 mm. Anterior of cephalosome very different from female, as it is not pointed and the rostrum is blunt. Antennule (A1) is twice geniculate, with a sheath just beyond the proximal elbow and a semicircular process on the first segment beyond the distal elbow.

Oncaea clevei

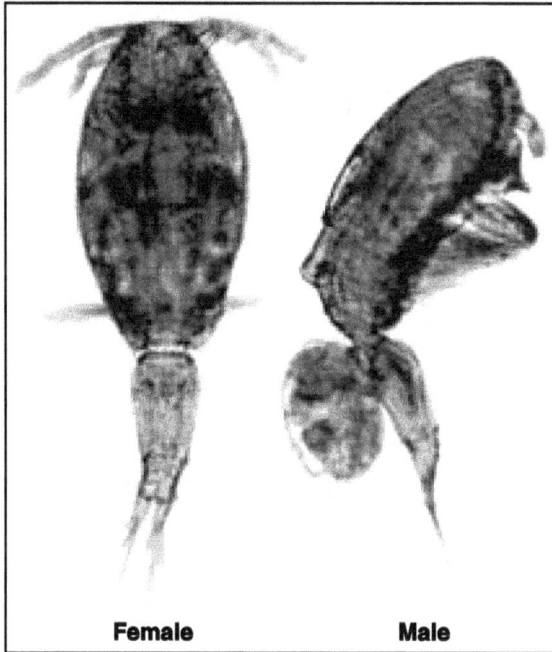

Female   **Male**

Female: Length: 0.65-0.75 mm. Prosome 1.8 urosome. Caudal rami length about 3 times width and approximately equal to sum of 2 preceding somites. Leg 5 with small free segment, 2 terminal setae of approximately equal length.

Male: Length 0.55-0.7 mm Prosome 1.7 times of urosome. Caudal rami 1.1-1.2 times anal somite.

Corycaeus dahli

Female: Length 0.97-1.10 mm. Prosome about 1.7 times as long as urosome. Urosome somites and caudal rami in the proportional lengths 30:18:38.

Male: Length 0.88-0.93 mm.Prosome about 1.4 times as long as urosome. Urosome somites and caudal rami in the proportional lengths 28:11:17. Genital somite oval in dorsal view; in lateral view ventral surface with a small median hook.

Copilia mirabilis

Female: Length 3.27-3.29 mm. Cephalosome quadrangular, widened posteriorly and as long as the rest of body excluding caudal rami. Ocular lenses separate by distance of approximately 1.5 their length. Caudal rami 1.5 times longer than urosome.

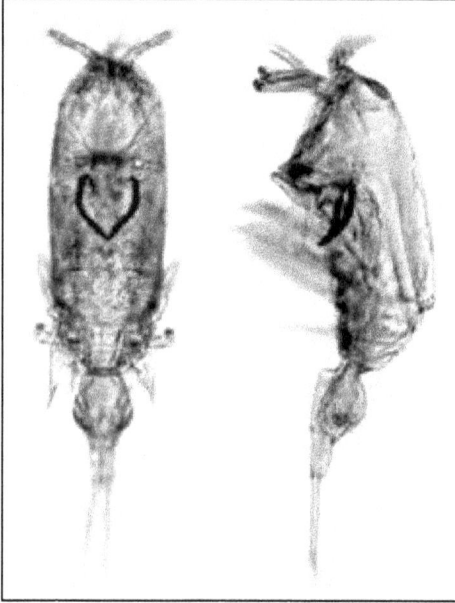

***Corycaeus dahli*–Female**

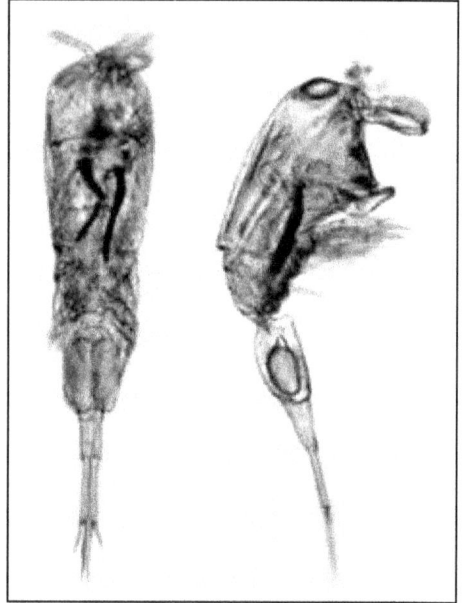

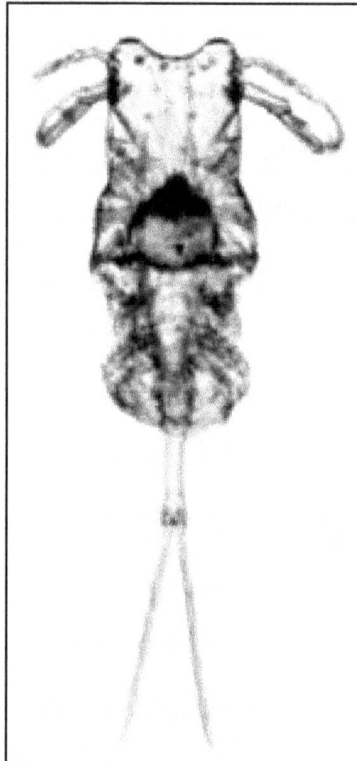

***Corycaeus dahli*–Male**

***Copilia mirabilis*–Female**

Sapphirina nigromaculata

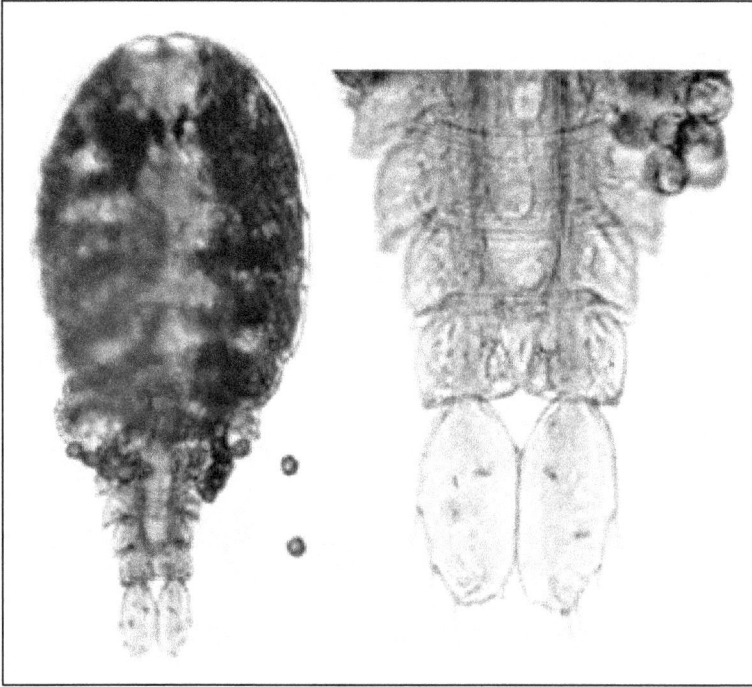

Female **Caudal Rami**

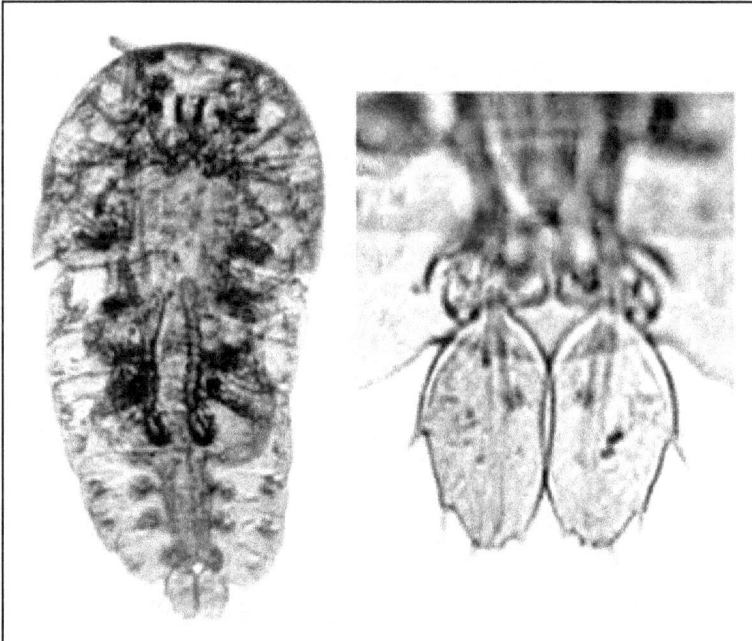

Male **Caudal Rami**

Female: Length 1.71-2.10 mm. Cuticular lenses visible in dorsal view. Prosome 1.8 times urosome. Pd1-3 tapered regularly backward. Caudal rami with 1 inner projection on distal border; length 2.3-2.5 times width.

Male: Length 1.93 mm Caudal rami with 1 projection on inner distal border, length 1.8 times width.

Suborder: Harpactocoida

Longipedia coronata

0.2 mm

Female: Head is devoid of eye lenses. Endopods of the second legs are greatly elongated. Inward expansion of the 5th legs is narrow, curved and pointed. Length, 0.8-1.3mm

Microsetella norvegica

Female: Length 0.65-0.72 mm. The last 3 urosomal somites with transverse rows of spinules, caudal rami outer apical seta short, the inner seta 1.2-1.4 times longer than the body length.

Male: Length 0.54-0.58 mm. Caudal rami inner seta 1.3-1.5 times longer than the body length.

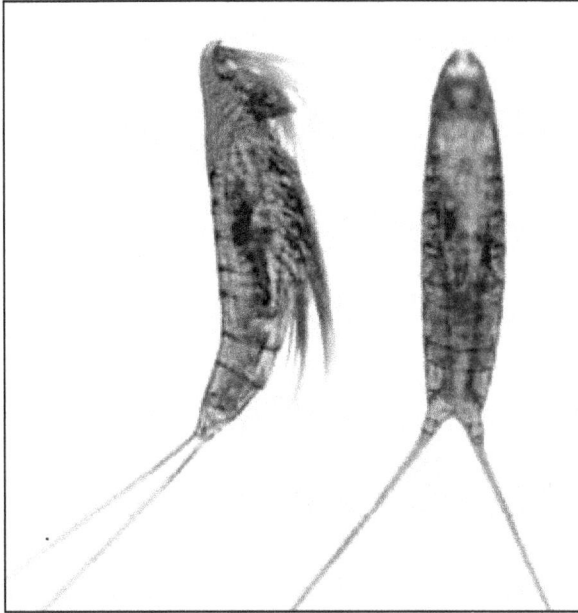
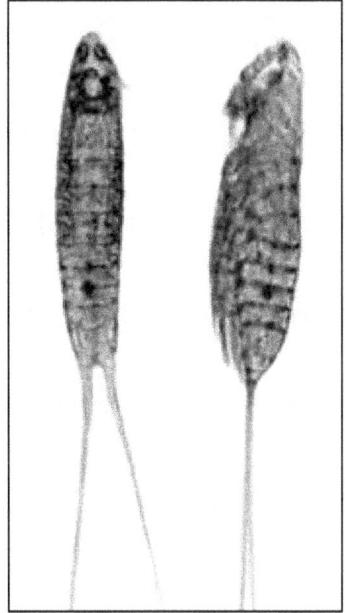

Female   **Male**

Macrosetella gracilis

Female: Length 1.22 mm. A1 relatively long, reaching Pd4. P5 basal expansion not reaching central part of distal segment, with 4 apical setae, the second inner seta plumose and the longest. Distal segment elongate, with 3 outer edge setae and 3 apical setae, of which 2 inner setae equal in length.

Male: Length 1.01 mm. A1 geniculate. P 5 basal expansion very short and tipped with 2 setae; distal segment elongate, with 3 apical and 1 outer setae.

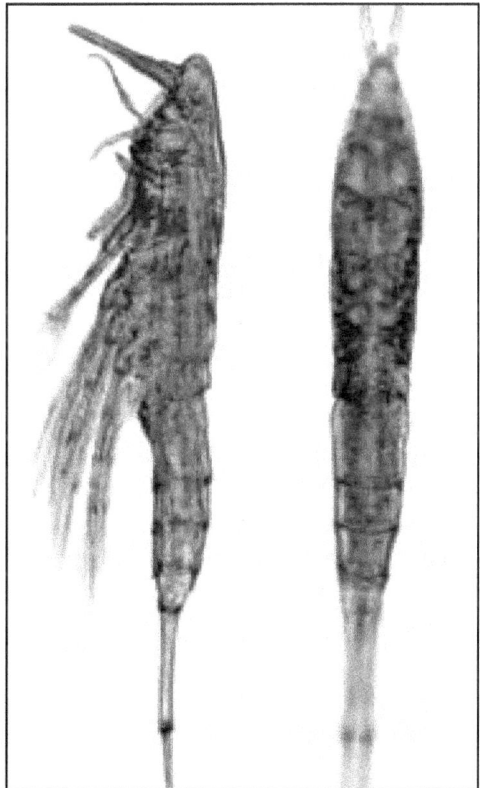

Female

Euterpina acutifrons

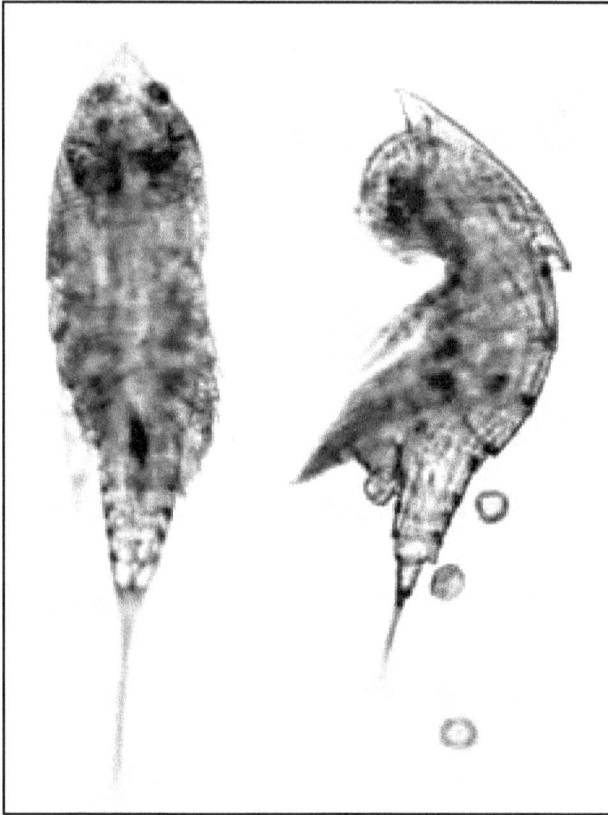

Female

Female: Length 0.68-0.72 mm. Body fusiform. Caudal rami and anal somite of approximately equal length.

Clytemnestra scutellata

Female: Length 0.86-0.92 mm. Genital somite longer than the following 2 somites combined; caudal rami 0.7 times as long as anal somite.

Male: Length 10.4 mm. It is easily identified by the rather stumpy shape of its body as well by the shape of the petasma.

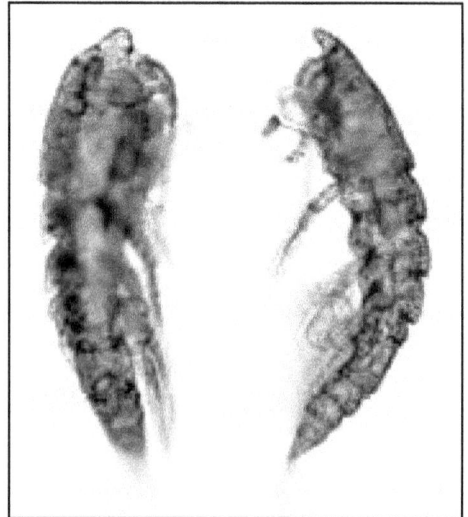

11.11. Mysids

Order: Mysida

Rhopalophthalmus sp.

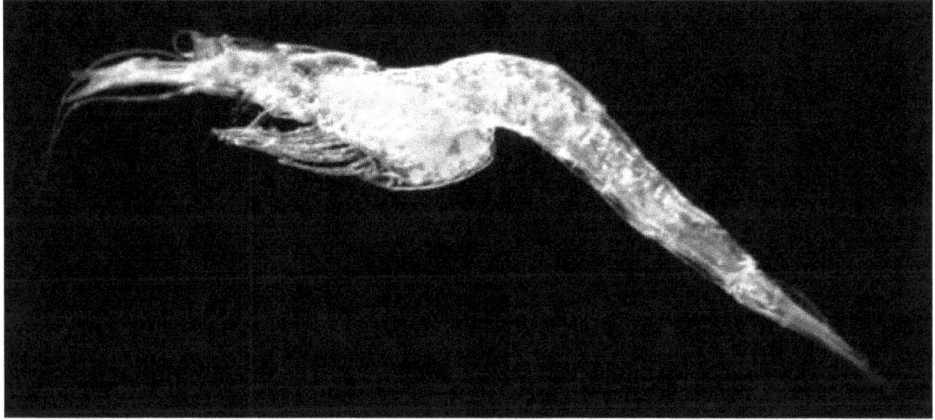

Female: Total length, 11.7 mm. Body slender. Carapace with dorso-median nodules; anterior margin lacking rostrum, postorbital spines prominent, keels prominent. Eyes stout, extending to distal end of the first segment of antennule, cornea occupying little more than half of eye and wider than stalk. Apical spines of telson are sub-equal. Number of spines on lateral margin of telson,14.

11.12. Decapods

Order: Decapoda

Lucifer hanseni

Description: This species is identified by the stumpy shape of its body, shape of the petasma, the sixth abdominal segment in males and by the comparative length of the outer spine of uropodal exopods in both sexes. Terminal portion of petasma acute and curved. Eyestalk less than half distance between bases of eye and labrum. Outer marginal spine of uropodal exopod not reaching lamellar part. Male's length, 10 mm.

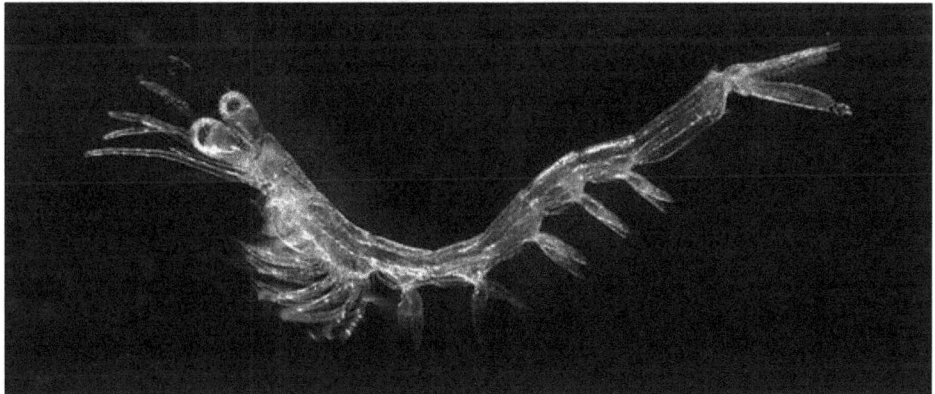

11.13. Molluscs

Phylum: Mollusca

Class: Gastropoda

Common Purple Snail

Order: Coenogastropoda

Janthina janthina

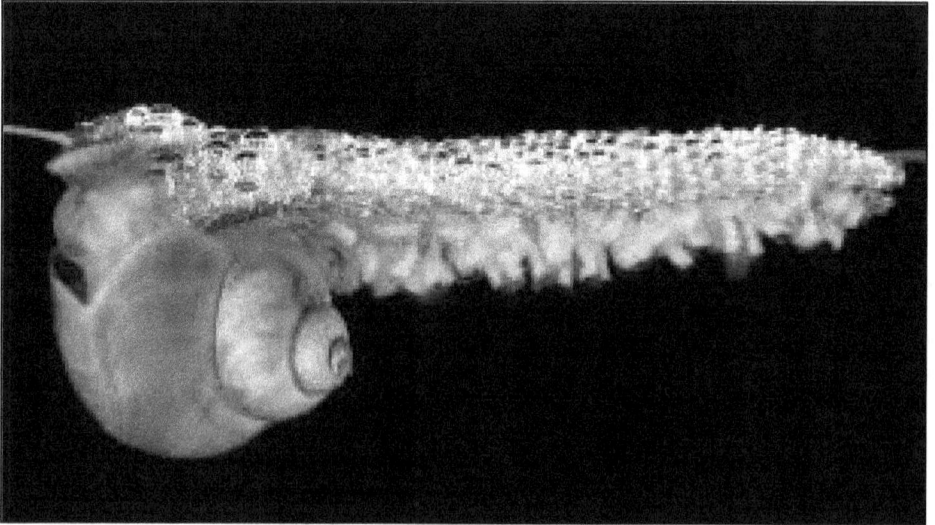

Description: This holoplanktonic species is popularly known as 'common purple snail'. These snails live as part of the pleuston (organisms living on or at the very surface of the water) because of their relatively large size. The adults do not swim and cannot create their rafts except at the surface where air bubbles are available. They trap air bubbles with a layer of clear chitin to maintain their positions at the surface of the ocean where they are prey upon the hydrozoans. In addition to the bubble raft, the shell is light and paper-thin to allow the animal to float upside down at the surface. The shell is almost smooth with a slightly depressed-globose shape. The colour of the shell is violet, with a paler upper surface. The height of the species shell is up to 38 mm and the maximum width is 40 mm. The animal has a large head on a very flexible neck. The eyes are small and are located at the base of its tentacles.

Pteropods (Sea butterflies)

Order: Thecosomata

Creseis acicula

Description: This species has a transparent uncoiled, shell with a smooth surface. The shell is extremely long and narrow, tube-shaped and not curved. The visceral

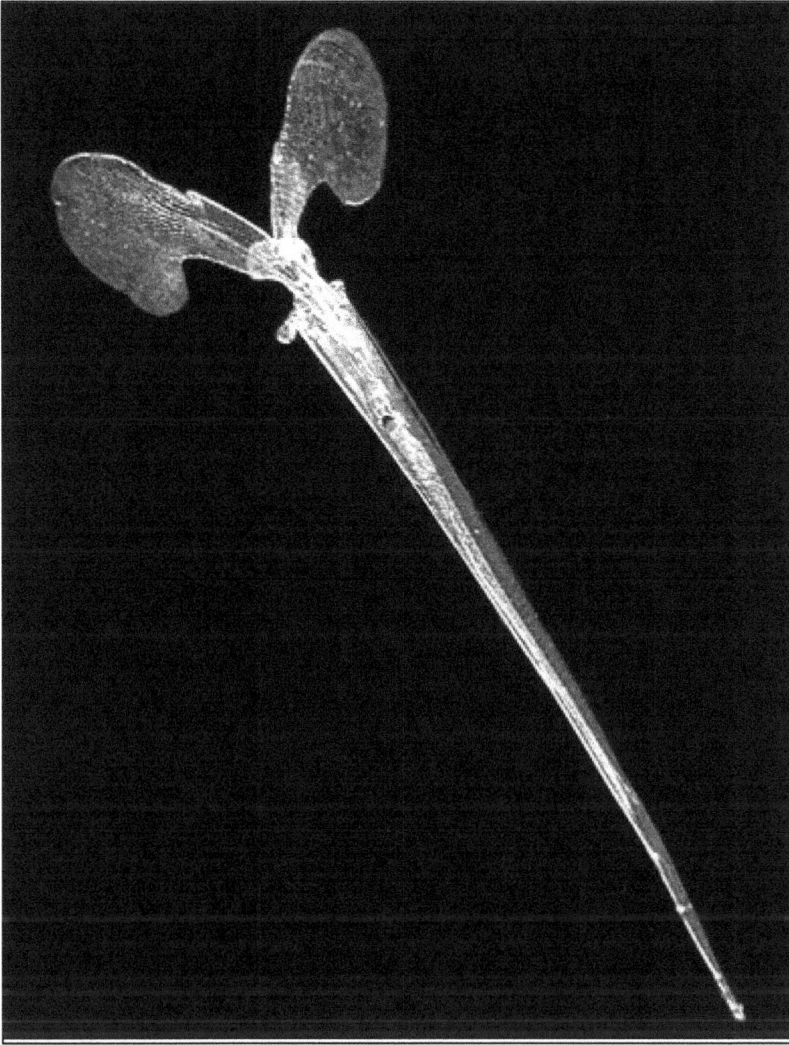

mass is seen through the shell. It is a clumsy swimmer feeding on phytoplankton and protozoans. Maximum length, 4 cm.

Cavolinia tridentata

Description: It is one of the larger uncoiled thecosomatous pteropods. It has a flat dorsal side with moderately developed ribs. There is no keel along the lateral sides, though the shell is very broad in relation to its length. The caudal spine is straight and relatively long. The ventral side is moderately vaulted. Shell sculpture has faint growth lines and a very faint transverse striation. It feeds on microzooplankton and phytoplankton and it is a mucus feeder. Length, 1 cm.

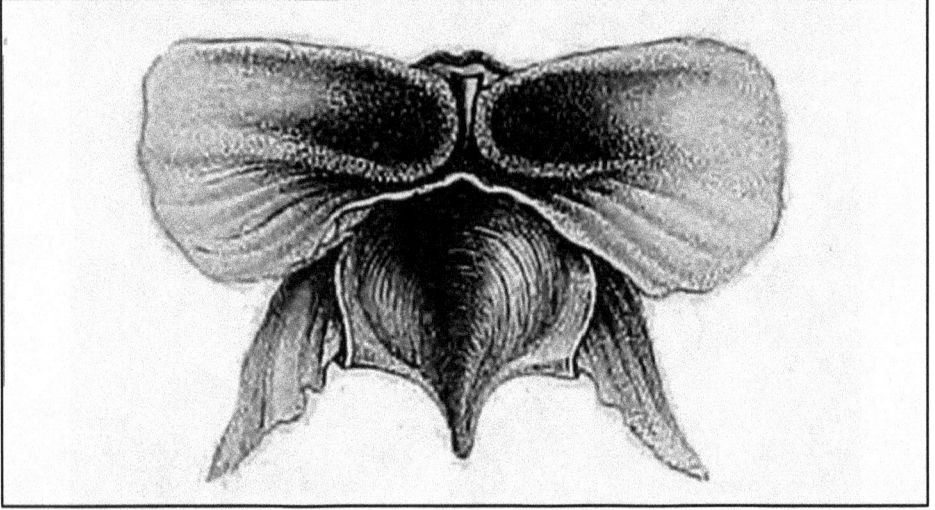

Limacina helicina

Description: This is a small, shelled, pelagic thecosomatous pteropod with a left-coiled shell of 0.6 cm diameter. The spire is very depressed. It has 6 colourless,

transparent whorls. There is no umbilical keel. Transverse striae are present. It feeds mainly on phytoplankton.

Limacina bulimoides

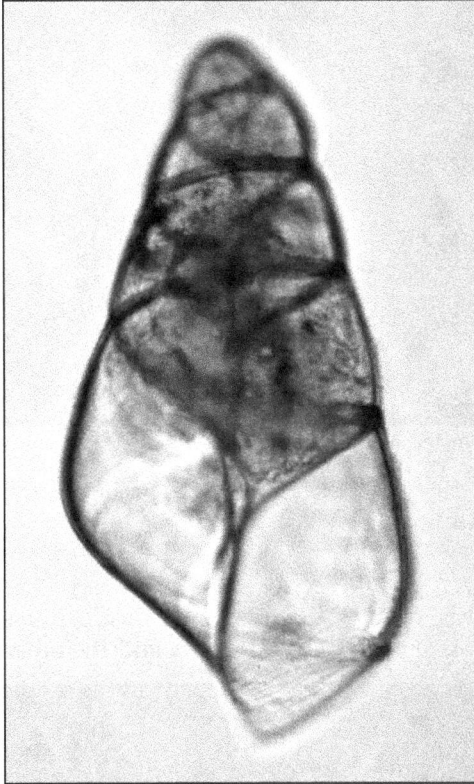

Description: The shell of this species has a spire with 5 whorls. Thicker parts of the shell near aperture brownish. Shell transparent. Shallow suture light-brown. Outer surface of the shell smooth. Small rostrum more pointed in older specimens. Shell height 1.3 mm, maximum diameter 0.6 mm.

Sea Slugs

Order: Gymnosomata

Clione limacina

Description: It is a common non-shelled, naked (gymnosomatous) pteropod and feeds exclusively on species of the shelled (thecosomatous) pteropod genus *Spiratella*. When it makes contact with its prey, it rapidly everts a set of six buccal cones, which are eversible tentacles, which grab the shell of *Spiratella* and turn it around until its shell opening is facing the mouth of *Clione*. Maximum length, 25mm.

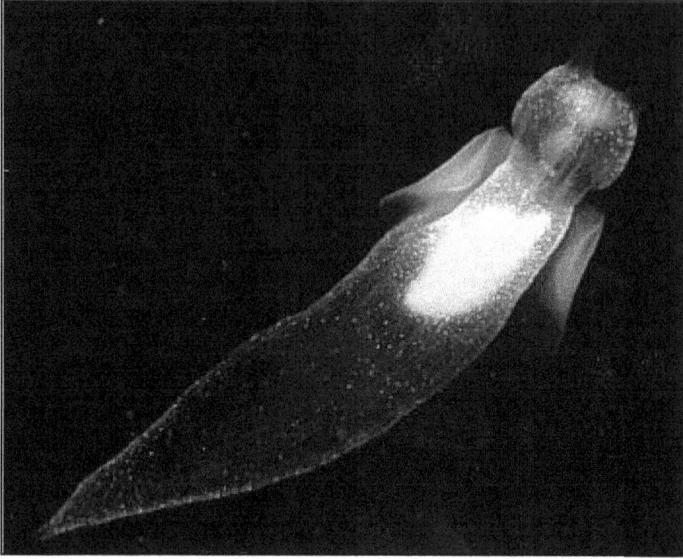

Nudibranchs

Order: Nudibranchia

Glaucus atlanticus

Description: This species always lives in close association with the siphonophores such as *Physalia, Velella, Porpita* and the other associated animals

including the "Violet snails" of the genus *Janthina.* All these animals float on the surface of the ocean and are normally carried by the currents and the winds. Onshore winds blow great fleets of them on to the beaches. It spends its life floating upside down in the water, partially bouyed by a gas bubble in its stomach. It feeds exclusively on *Physalia.* Maximum length, 3cm.

11.14. Protochordates

Planktonic protochordates are normally seen both in epipelagic and mesopelagic ecosystems. The Larvaceans or "salps," secrete a mucous "house" and live inside where they actively pump water and feed on smaller planktonic organisms. When disturbed, they swim freely away from the mucous house.

Larvaceans

Phylum: Chordata

Subphylum: Urochordata

Class: Larvacea

Order: Appendicularia (Copelata)

Oikopleura dioica

Description: Trunk compact. Buccal glands spherical and small. Dioecious. Testis or ovary adjoining the coil of the gut hemispherical. Tail with 2 spindle-shaped

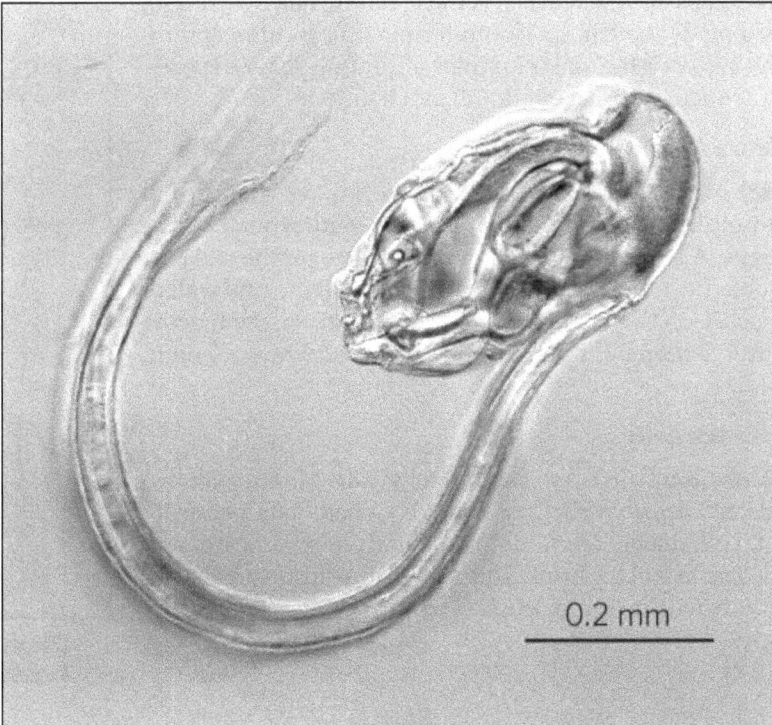

0.2 mm

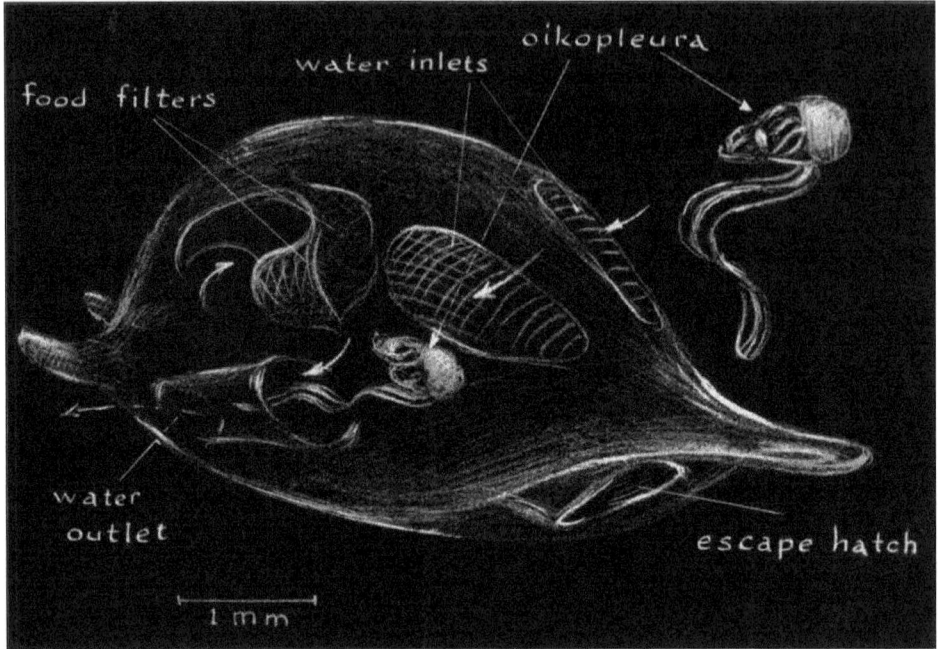

House of *Oikopleura*

subchordal cells arranged in a line, musculature narrow. Tail six times as long as the trunk. Mouth terminal or located antero-dorsally. *O. dioica* is the most eurythermal and euryhaline species living in a mucus house. Length, 1.5 mm long.

Oikopleura longicauda

Description: Trunk of this species compact. Postcardial caecum pointing dorsally, adjoining the genital wall of the oesophagus. A long velum extending dorsally over the oral part of the trunk. Rectum extending scarcely beyond the oral wall of the stomach. Gonads adjoining the coil of the gut, and embracing it laterally when ripe. Tail musculature broad. Maximum length, 1.4 mm.

Frittilaria borealis

Description: Trunk of this species elongate. Mouth with two lateral plates. Digestive tract axis oblique. Gonads symmetrically arranged. Testis elongate and simple, behind one spherical ovary, on the median axis of the trunk. Tail musculature from moderately broad to narrow. Maximum length, 1.5 mm.

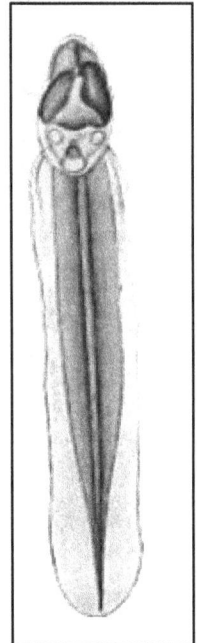

Oikopleura longicauda

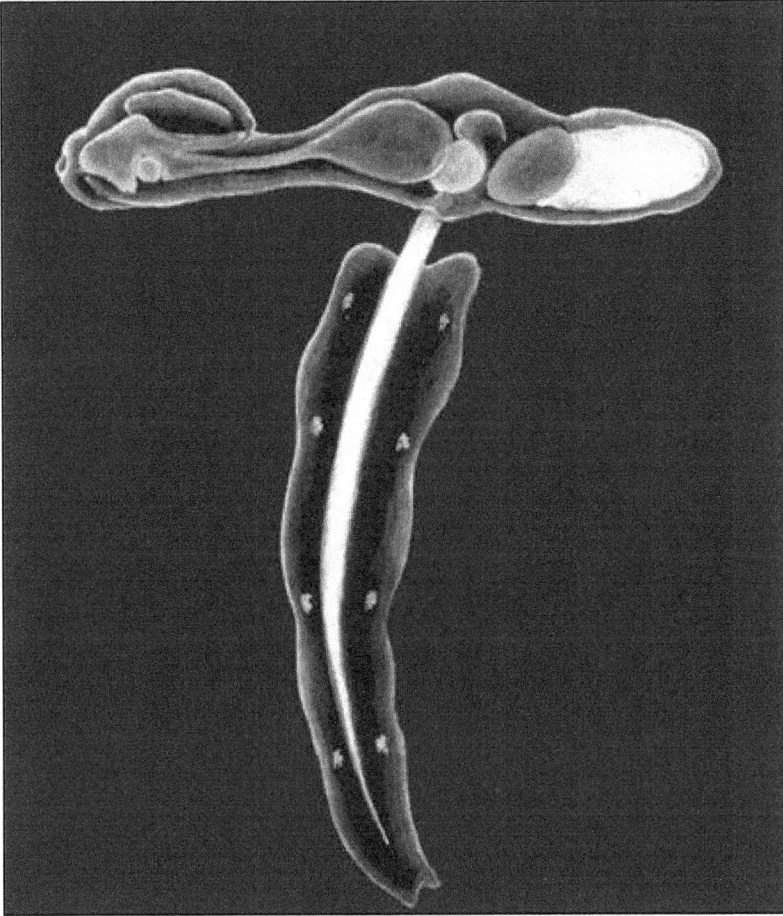

Frittilaria borealis

Thaliaceans (Pyrosomes, Salps and Dolioloids)

Class: Thaliacea

Order: Pyrosomida

Pyrosoma atlanticum

Description: It is a colonial tunicate. The colony of this species is cylindrical. The constituent zooids form a rigid tube, which may be pale pink, yellowish, or bluish. One end of the tube is narrower and is closed while the other is open and has a strong diaphragm. The outer surface or test is gelatinised and dimpled with backward pointing, blunt processes. The individual zooids are up to 9 mm long and possess a broad, rounded branchial sac with gill slits. Along the side of the branchial sac, runs an endostyle, which produces mucus filters. Water is sent through the gill slits into the centre of the cylinder by cilia pulsating rhythmically. Plankton and other food

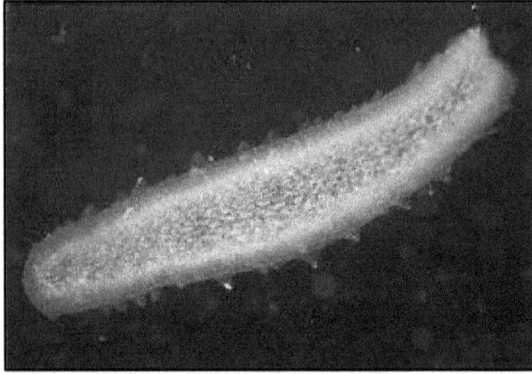

organisms are caught in mucus filters in the processes as the colony is propelled through the water. It is bioluminescent and may generate a brilliant blue-green light when stimulated. It grows up to 60 cm long and 6 cm wide.

Salps

Order: Desmomyaria (Salpidae)

Salpa (Thalia) dmocratica

Description: It is aggregate form with anterior end of body rounded, the posterior end is somewhat fusiform. The test is bluntly pentagonal. M1-M3 and M4-M5 (muscle bands) are fused mid-dorsally; both groups of muscle bands are separated in lateral view. The nucleus has a posterior projection. The test is pentagonal posteriorly. The number of muscle fibres in M1-M5 is 16. The atrial opening is not central and there is

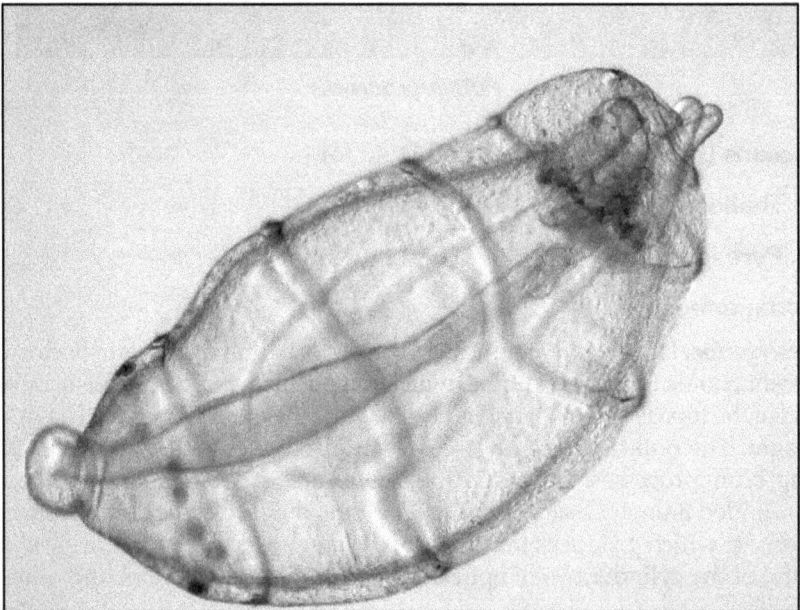

a posterior lateral protuberance on one side only. Solitary form with smooth test, atrial palps simple and straight; lateral projections small, medio-ventral projections are small and of unequal length, the anterior ones are the smallest. All test projections are echinate. M1-M4 is continuous dorsally and ventrally; M5 is narrowly interrupted ventrally. M2-M4 and M5-M6 are converged mid-dorsally over a short distance only. The stomach partly extends into the middle posterior projection. Maximum length of aggregate zooids 18.0 mm and solitary zooids 11.5 mm.

Doliolids

Order: Cyclomyaria (Doliolida)

Dolioletta gegenbauri

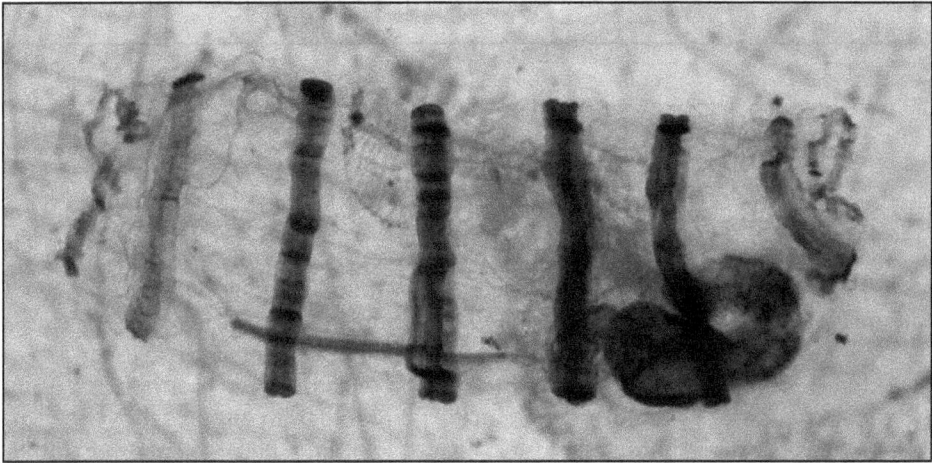

Description: This is a small, transparent, gelatinous species growing up to 1 cm. It has a complex life cycle and exists in several forms of which the gonozooid, or mature zooid with gonads, is often seen. It is roughly cylindrical with a siphon at both of the flat ends, and has eight bands of muscle arranged like hoops round a barrel. The U-shaped gut and other organs may be seen through the test which is pierced by 10 to 40 gill slits. The gonozooid is hermaphrodite and the eggs are fertilized by the sperm of another individual. These develop into oozooids which have no reproductive organs. They have 9 bands of muscle and are known as "nurses" as they develop a tail of zooids produced asexually.

Chapter 12
Identification of Zooplankton– Meroplankton

Meroplankton are organisms that are planktonic for only a part of their life cycles, usually the larval stages. Examples of meroplankton include the larvae of sea urchins, sea stars, crustaceans, marine worms, some marine gastropods and most fish. While living in the plankton, meroplankton either feed on other members of the plankton, or depend on the yolk they have retained from the egg they hatched from. Larvae spend varying amounts of time in the plankton, from minutes to over a year. After a period of time in the plankton, the meroplankton become either nekton or adopt a benthic life on the seafloor.

12.1. Larvae of Cnidarians

Most cnidarians possess a "larval" form that develops from the fertilized egg, and swims until it finds a place to anchor and metamorphose into a polyp. The typical cnidarian "larva" is a non-feeding small, biradial, planula. Among the cnidarians, the best studied genus is *Aurelia.* There are separate sexes, male and female. Reproduction begins when the male releases sperm through its mouth into the surrounding water. These swim to the female where they enter her central oral cavity to reach the eggs. Once fertilized, the zygotes emerge onto the oral arms to develop for a time, becoming larvae which settle on the bottom of the ocean. The resulting polyp begins to bud asexually, releasing free-swimming medusae which go on to develop into adults.

Phylum: Coelenterata

Class: Scyphozoa

Larvae of *Aurelia aurita*

The scyphozoan medusa possesses alternation of generation in its life cycle. The different stages are given below:

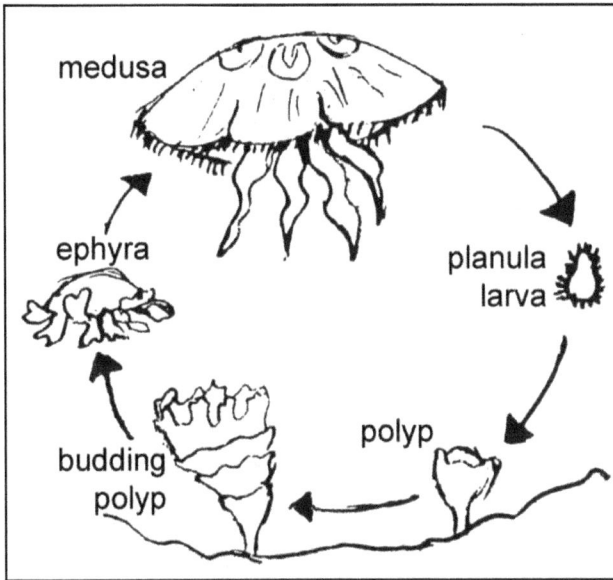

Life Cycle of *Aurelia* sp.

Planula: It is elongated and radially symmetrical with distinct anterior and posterior ends. Gastrovascular cavity and mouth are absent.

Scyphistoma: It is distinguished by the presence of 20 tentacles, 4 septal funnels and 4 septa which are visible through the open mouth. During the start of strobilation, a constriction (first circular furrow) develops at its distal end.

Strobila: During strobilation, the scyphistoma develops one or several transverse constrictions and becomes a monodisc and polydisc strobila. Asexual phase of metagenesis begins now.

Ephyra: An young ephyra budded off from the strobila is characterized by the presence of 4 perradial and 4 interradial marginal lappets and each of 2 wing lappets with 4 interradial lappet pouches and 8 marginal pouches. Through a number of transformations, the ephyra gradually changes to an young scyphomedusa which on reaching the sexual maturity once again starts the alternation of generation.

12.2. Larvae of Planktonic Flatworms

Most of the species in the order Polycladida (Phylum: Platyhelminthes, Class: Tubellaria) produce larvae within direct development. In these species, a free-swimming Müller's larva is normally formed.

Phylum: Platyhelminthes

Class: Turbellaria

Müller's Larva

Description: This larva is small, helmet-shaped, and possesses eight ventrally directed ciliated lobes, which help in its propulsion. This larva has ciliary bands on

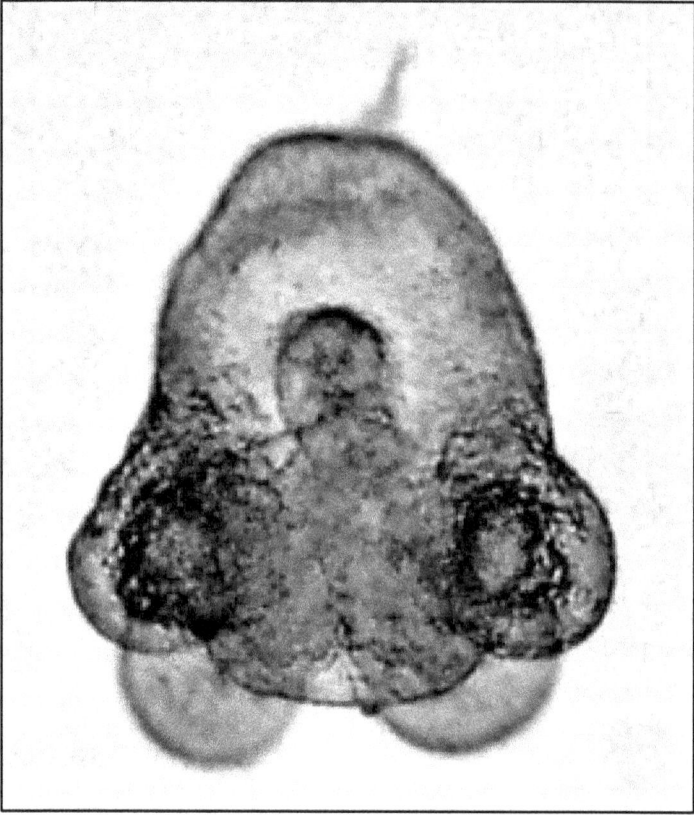

swimming arms and apical tuft of cilia on the anterior side. Mouth is ventral and anus is absent. It is planktonic for several days before it settles to the bottom with its oral surface downward and metamorphoses into a juvenile flatworm. Size of this larva is 0.15-1.8 mm.

12.3. Larvae of Nemertine Worms

Nemertines are commonly called ribbon-worms or proboscis worms. Their pilidium larvae which are free-swimming develop from initial simple ciliated larvae.

Phylum: Nemertinea

Pilidium Larva

Description: It has a dome-shaped body with a tuft of sensory cilia sprouting from the top of the apical plate. The lower body is formed by 4 lobes, single anterior and posterior lobes which are ciliated and are used in locomotion. The size of pilidium larvae is 0.5-0.8 mm (from apical tuft to distal edge of the lateral lobe).

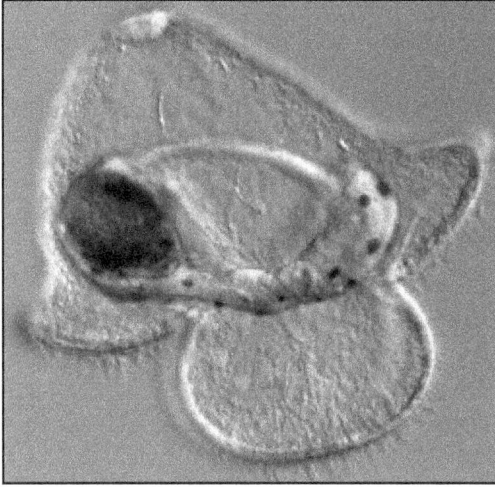

12.4. Larvae of Brachiopods

Brachiopods or 'lamp shells' are sedentary filter-feeding animals which resemble the bivalved molluscs to some extent. The larvae of these animals swim as plankton for months and are like miniature adults, with valves, mantle lobes, a pedicle and a small lophophore, which is used for both feeding and swimming.

Phylum: Brachiopoda

Lingula Larva

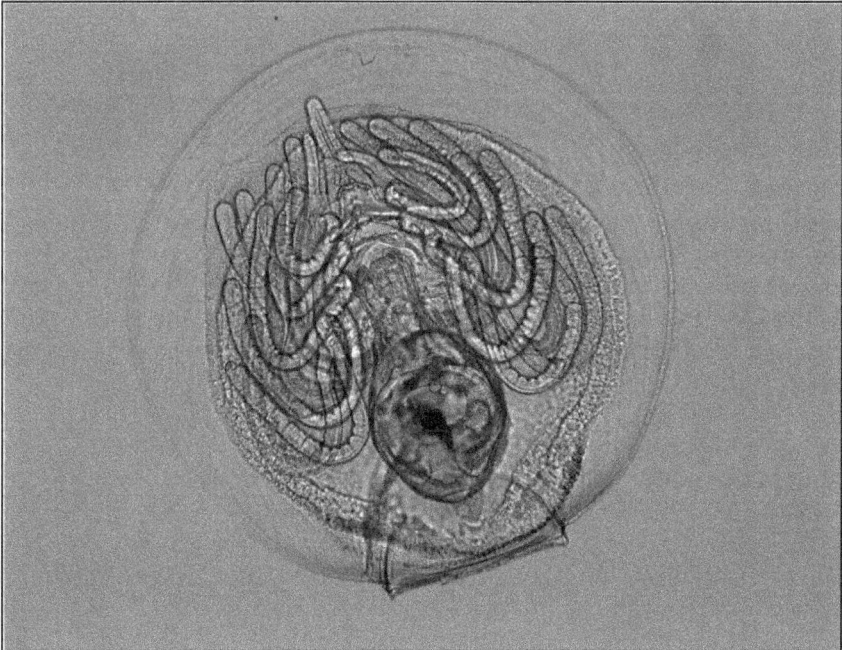

Description: The chief characteristics of the Lingula larva are dorsal and ventral shell valves underlain by mantle tissue, a complete gut, a developing pedicle, and a lophophore. Each shell valve is divided into a posterior semicircular embryonic shell or protegulum and a larger, circular to ovoid larval shell, which is added as larvae grow in the plankton. The lophophore consists of pairs of cirri that bud off sequentially on either side of the median tentacle, so that the youngest tentacle pair is most anterior. The median tentacle is a sensory organ, a characteristic feature of lingual larva. The anterior lip of the mouth is called as epistome.

12.5. Larvae of Phoronids

The phylum Phoronida is an obscure group of sessile, benthic worm-like animals which are found largely attached to submerged pier pilings, boat docks, and other underwater structures that offer exposure to non-turbulent water flow. The actinotrocha larva of phoronids is extremely rare in the open coast, nearshore plankton hauls.

Phylum: Phoronida

Actinotrocha Larva

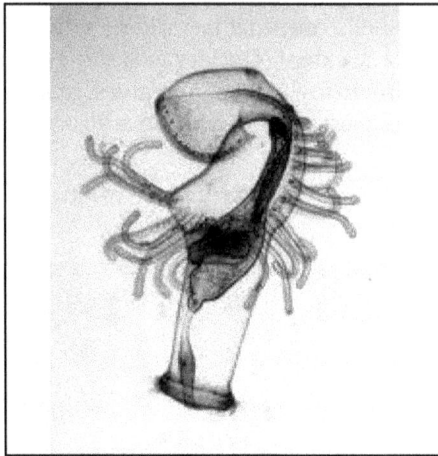

Description: The actinotrocha larva has an anterior preoral lobe on which is located the nervous ganglion (on the apical area), a tentacular ridge, a pair of protonephridia and posteriorly a ciliated ring around the anus. The larvae undergo a planktotrophic development from 2–3 weeks and settle after about 20 days. Metamorphosis is very fast, and a slender young phoronid develops in less than 30 minutes.

12.6. Larvae of Bryozoans

Bryozoans are benthic colonial animals which attach and spread over rocks, kelp, shells and other marine structures. They are tiny worm-like creatures living inside boxes and are barely visible to the naked eye. Bryozoans reproduce, develop and distribute themselves using their triangularshaped cyphonautes larvae. Cyphonautes shells are transparent, double and the open edge is fringed with active cilia. These larvae drift in the currents until ready to settle out to the bottom and take up a sessile benthic life.

Phylum: Ectoprocta

Cyphonautes Larva

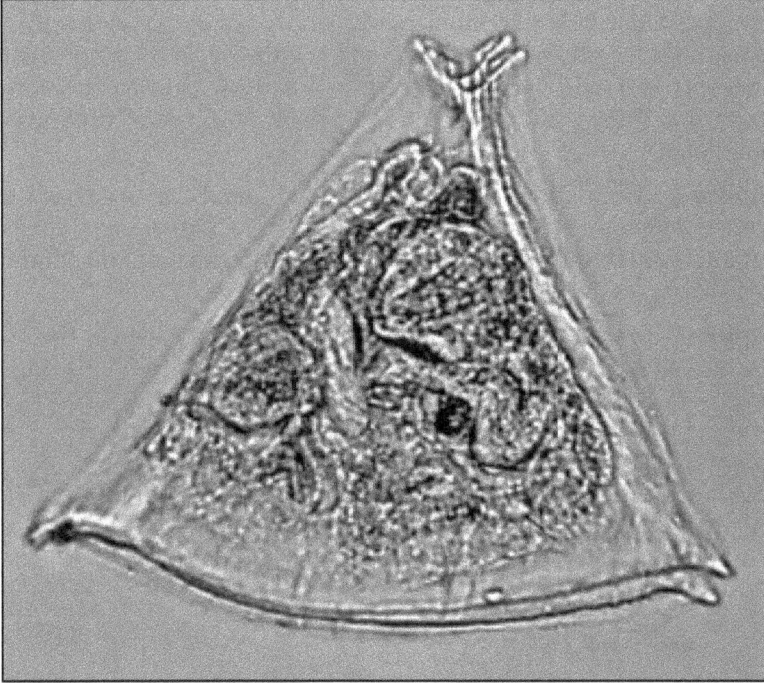

Description: Bryozoan larvae are 0.2-0.6 mm in length, vary in form, but all have a band of cilia round the body which enables them to swim, a tuft of cilia at the top and an adhesive sac that everts and anchors them when they settle on the surface. Some species of class Gymnolaemata produce cyphonautes larvae which possess little yolk but a well-developed mouth and gut, and live as plankton for a considerable time before settlement. These larvae have triangular shells of chitin, with one corner at the top and the base open, forming a hood around the downward-facing mouth.

12.7. Larvae of Crustaceans

Crustaceans begin life as a developing embryo inside an egg which is being carried along with hundreds or thousands of other eggs. Some species, on the other hand, shed their eggs into the current. The first stage after hatching from the egg is the nauplius stage. There may be several naupliar stages which are metamorphosed into a second larval stage known as the cypris larva. The cypris may develop into a zoea, in crabs, or a phyllosoma, in the case of the spiny lobster. The next stage is the mysiid shrimp, a big crustacean.

Class: Crustacea

Order: Copepoda

Larvae of Calanoid Copepod, *Acaria* **sp.**

Naupliar Stages

During postembryonic development, six naupliar stages are seen in this species. Average naupliar length is about 2.1 times its width. The body is not significantly curved laterally. All naupliar stages are oval anteriorly, and narrow toward the caudal armature. There is an anteroventral pigment spot, generally red, also known as the naupliar eye.

Nauplius I: Length range is 0.081-0.113 mm. Caudal armature has 2 terminal sensory setae Antennule, antenna, and mandible are rudimentary. After the first molt, the nauplii enlarge slightly, and the antennule, antenna, and mandible become more specialized.

Nauplius II: Length range 0.11 1-0.123 mm. Body is egg-shaped. The 2 terminal sensory setae of the caudal armature are longer than in the previous stage; one is

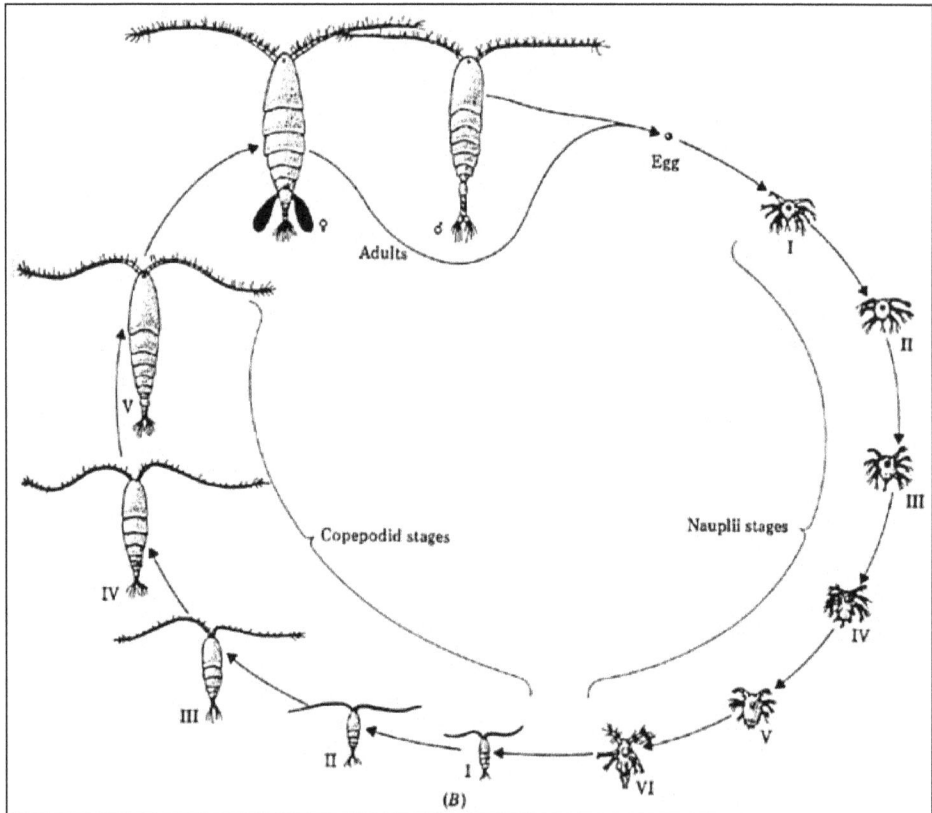

Life Cycle of a Calanoid Copepod.

ventral, the other dorsal. The labrum is oval. Posteroventrally the body has a transverse row of fine setae.

Nauplius III: Length range 0.127-0.147 mm. The body remains egg-shaped. The caudal armature now consists of 2 ventral spines with slightly toothed margins (saw-type). Posteroventrally the body has 2 transverse rows of setae.

Nauplius IV: Length range 0.157-0.176 mm. Body is oval. The caudal armature is larger than in naupliar stage 111, and has 2 terminal spines with toothed margins, 2 anterior sensory setae and a new pair of ventral spines.

Nauplius V: Length range 0.189-0.208 mm. The caudal armature is a small transverse row of small setae: 4 spines remain as in naupliar stage IV (2 ventral strong and small, 2 large dorsal, terminally toothed); there are also 2 sensorial setae. Labrum is oval in shape.

Nauplius VI: Length range 0.218-0.240 mm. Until this last naupliar stage the body remains oval. In lateral view a clear differentiation of 5 segments is seen. Labrum is oval and the caudal armature is the same as in nauplius stage V.

Copepodid and Adult Stages

The final naupliar stage metamorphoses into the first of six copepodids. When the individual reaches copepodid stage I, it is a miniature adult, except that it has only two pairs of functional swimming legs. A new pair of legs is added in each successive molt until copepodid IV. From that stage until the adult form (copepodid VI), no new swimming legs are added; instead, the sexually modified fifth pair of legs develops in the adult form. During the copepodid stages, males increase 2 times in average body length to reach the adult stage; females increase 2.5 times. From copepodid IV on, the sex of each individual is evident. The most important characters of the copepodid stages of this species are given below:

Copepodid I: length range 0.347-0.443 mm. The cephalosome occupies about 60 per cent of the prosome length. The metasome has 4 segments: the first two have functional swimming legs; the third, in some cases, has biramous buds of the third pair of swimming legs; and the fourth lacks appendages. The urosome has one segment. In the caudal furca each ramus is longer than it is wide and has 4 plumose setae. The rostrum bears two filaments.

Copepodid II: Length range 0.424-0.500 mm. The cephalosome occupies 54 per cent of prosome length; it has 2 rostral filaments. The metasome has 4 segments, of which the first 3 have functional swimming legs and the fourth lacks appendages. The urosome has 2 segments. Each ramus of caudal furca has 5 plumose setae.

Copepodid III: Length range 0.519-0.635 mm. The cephalosome occupies nearly 52 per cent of the total length of the prosome. Metasome has 4 segments, each with a pair of functional swimming legs. The caudal furca has 6 setae; the same number occurs in the following copepodid stages. Forehead has 2 rostral filaments.

Copepodid IV, Male: Length range 0.597-0.674 mm. From this stage on, sexes are easily distinguishable. The cephalosome occupies nearly 51 per cent of the total length of the prosome. Each of the 4 metasomal segments has a pair of swimming legs; the posteriormost segment also has the two-jointed fifth pair of legs, which is symmetrical. The urosome has 3 abdominal segments. The caudal furca has 2 furcal rami with 6 plumose setae each. The species has two rostral filaments.

Copepodid IV, Female: Length range 0.635-0.751 mm. Female has the same general characteristics as the corresponding male stage, except that the urosome has 2 segments. The fifth pair of legs is symmetrical.

Copepod V, Male: Length range 0.693-0.809 mm. The cephalosome occupies nearly 46 per cent of the prosome's length. The metasome has 4 segments. Each one has a pair of swimming legs. The urosome has four segments. Each furcal ramus has 6 plumose setae.

Copepodid V, Female: Length range 0.712-0.924 mm. The female is larger than the corresponding stage of male. The cephalosome is approximately 46 per cent as long as the prosome, and has 2 rostral filaments. In general, the female has the same characteristics as the male, except that urosome has 3 segments, of which the first is larger. The fifth swimming legs are symmetrical.

Copepodid VI, Male: Length range 0.770-0.924 mm. The prosome has 6 segments. The cephalosome is approximately 50 per cent as long as the prosome; it has 2 rostral filaments. The metasome has 4 segments. The urosome has 5 abdominal segments. Furcal rami have 6 plumose setae (the innermost is the smallest). The fifth legs are uniramous and asymmetrical.

Copepodid VI, Female: Length range 0.924-1.097 mm. The female is longer than the same male stage (average ratio 1:0.82). The cephalosome is nearly 46 per cent as long as the prosome; it has 2 rostral filaments. This stage has the same general characteristics as the corresponding male stage, except that the urosome has 3 segments: the first one (genital) is longer than it is wide, and the third one (anal) is shorter and lacks spines and setae. The caudal furcal rami are longer than they are wide. The fifth legs are uniramous and symmetrical.

Larvae of Penaeid Prawn, *Penaeus* sp.

The penaeid prawns have a complete development from a nauplius stage unlike other crustaceans which hatch at a later stage either as a zoea or in an even more advanced stage.

Naupliar Stages

Nauplius I: In this stage, an ocellus is present at the anterior median region of the body. A pair of dorsally curved caudal setae are present at the anterior end. Among the 3 pairs of appendages, the first pair is uniramous and the other two pairs are biramous. The last pair is however is shorter than the other appendages. The setae are non-plumose. Size: Length, 0.30 mm.

Nauplius II: It resembles largely the naulius I. The setae are plumose and the furcal setae show a faint demarcation at the proximal end. Size: Length, 0.31mm.

Nauplius III: The furcal lobes bear 3 setae each. Of these setae the innermost is very small and ventrally placed. Size: Length, 0.31mm.

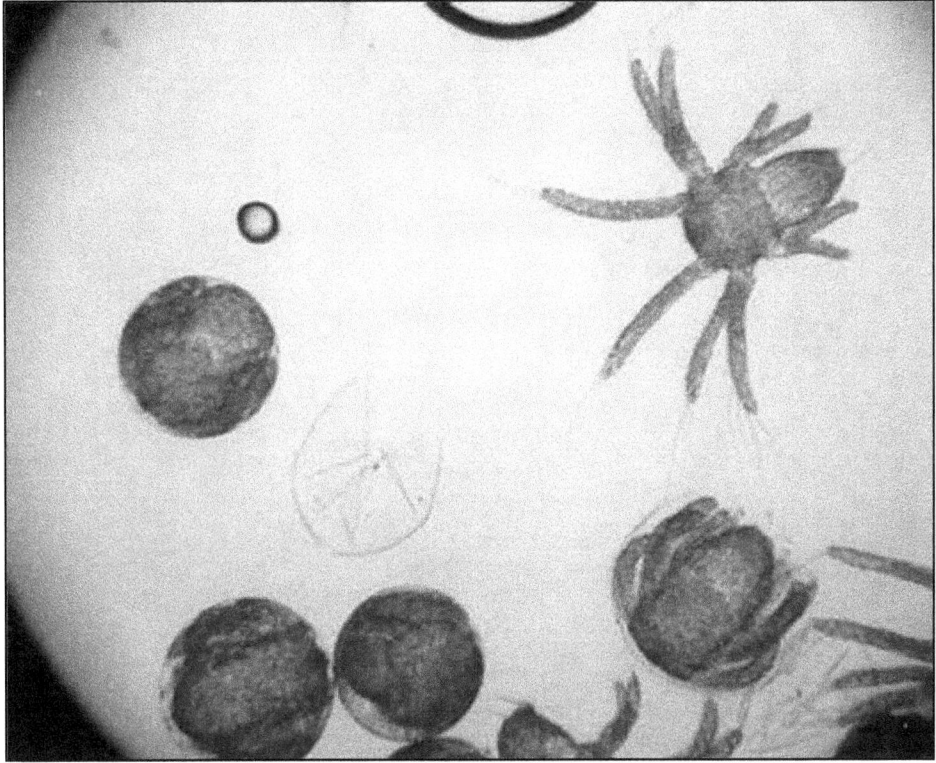

Eggs and Recently Hatched Nauplii.

Nauplius IV: The furcal lobes are very distinct with 4 setae each. The outer most setae are the smallest and are dorsally placed. Size: Length, 0.36mm.

Nauplius V: The furcal lobes are well developed with 6 setae each. The outermost are minute and are dorsally placed. Size: length, 0.38mm.

Nauplius VI: The body of the larva is more elongated with more or less distinct frontal organs and carapace. The furcal lobes bear 7 setae each. Size: Length, 0.48mm.

Protozoeal Stages

Protozoea I: The carapace is anteriorly rounded with a median notch. The frontal organs develop as rounded protuberances. The ocellus however, persists. The body is divisible into the carapace, the thorax and the abdomen which is unsegmented. The caudal furcal lobes bear 7 setae each. Size: Length, 0.88mm.

Protozoea II: This stage is characterized by the presence of a well developed and curved rostrum, bifurcated supraorbital spines and stalked compound eyes. The frontal organs are absent. Size: Length, 1.5mm.

Protozoea III: In this stage, the supraorbital spines are not bifurcated. The telson is distinct and is demarcated from the 6th abdominal segment by an articulating joint. Abdominal segments 1-5 bear a dorsomedian spineon the posterior border. The 5th and the 6th abdominal segments each have pair of posterolateral spines. The

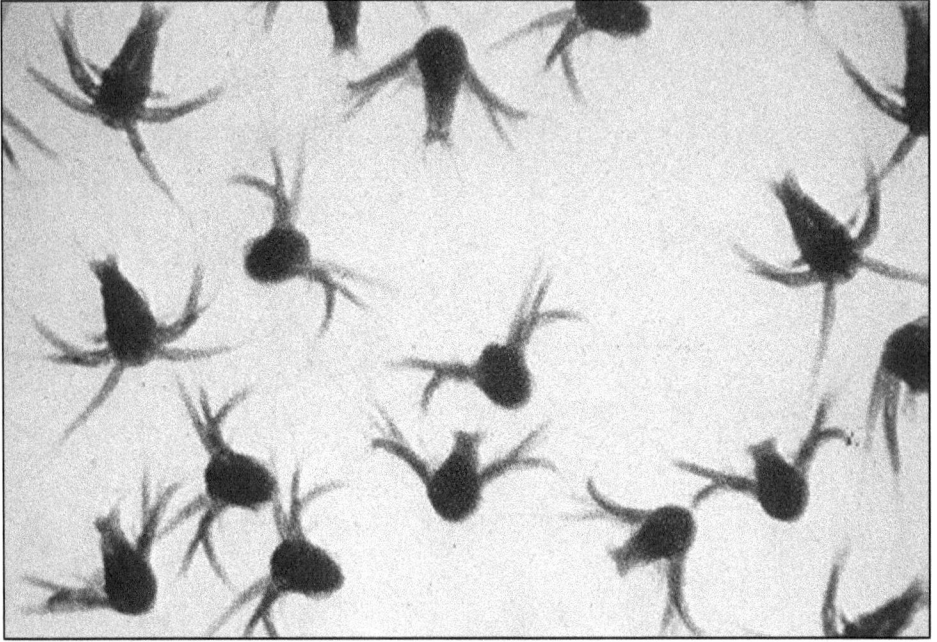

Nauplius II

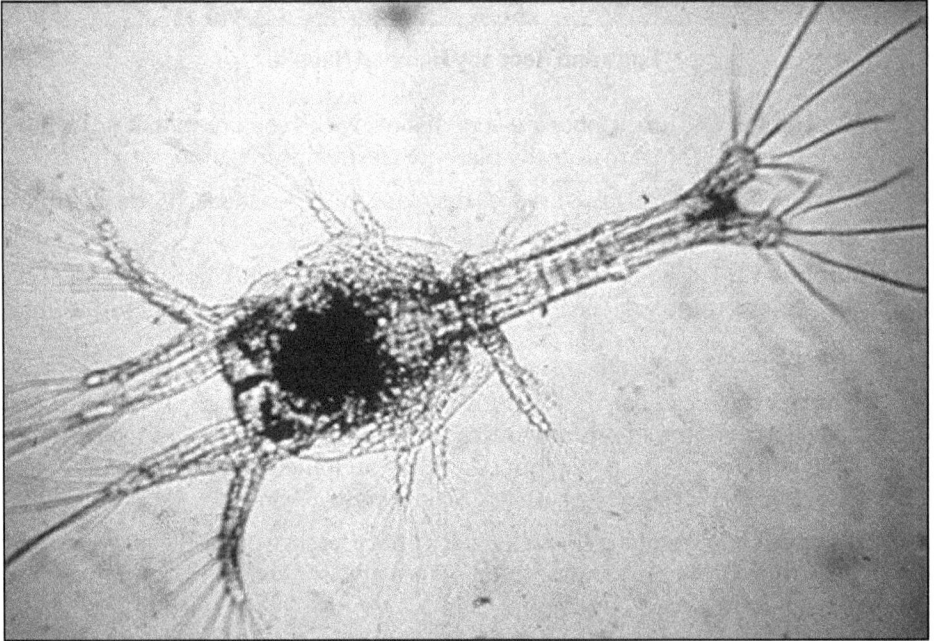

Protozoea

6th segment is devoid of a posteromedian dorsal spine but has a pair of ventrolateral spines. The caudal furcae bear 8 setae each. Size: length, 2.7mm.

Mysis Stages

Mysis I: It is more or less shrimp-like with a long rostrum which is curved extending beyond the eye. The rostral spies are absent. A prominent supraorbital spine and a small spine are present at the anteroventral angle of the carapace. The hepatic spine is well developed. The carapace completely covers the thoracic region and the thoracic appendages are well developed. The 5th and the 6th abdominal segments bear posterolateral spines. The abdominal segments may also have one dorsal spine each on their posterior margin. The telson is broader distally with a median notch and each of its lobes bears 2 lateral and 6 terminal setae. The cleft of the telson extends to the level of half-way between the origin of the outermost and penultimate pair of setae. Size: Length, 3.4mm.

Mysis II: It largely resembles mysis I and is characterized by the presence of a spine on the scaphocerite and an unsegmented pleopod bud. The cleft of the telson extends to the level of the origin of the penultimate pair of the lateral telsonic setae. Size: Length, 3.5mm.

Mysis III: It is characterized by the presence of a 2 segmented pleopod bud. The telson is long and rectangular carrying 6 distal and 2 lateral setae oneach side. The cleft of the telson extends to the level of the origin of the 3rd pair of setae. Size: Length, 3.9mm.

Postlarva: This stage is characterized by the presence of a rostrum which bears 1-2 dorsal spines. The supraorbial, hepatic and pterygostomial pines are present. The 4th, 5th and 6th abdominal segments are provided with median dorsal spines. The 5th and 6th segments may also have lateral spines. The pleopods are well developed and setose. The telson is rectangular carrying 3 pairs of lateral and 5 pairs of terminal setae. The median notch is absent in the telson.

Larvae of Barnacles

Order: Cirripedia

Nauplius

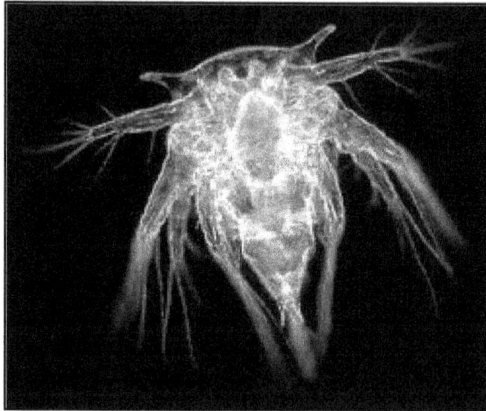

Description: The nauplius stage of barnacles has an unsegmented, pyriform body, enclosed in a carapace which has long caudal and dorsal spines. There are three pairs of limbs, in which the first pair (representing the antennas) is undivided, while the two hinder pairs are bifid, and all carry natatory bristles. A very large labrum is seen in front of the mouth and a well-developed alimentary tube terminates by a distinct anus at the root of the caudal spine. A simple central eye is present initially but in the late nauplius, additional two compound lateral eyes are developed. The nauplius develops into a cypris stage.

Cypris

Description: After 5-6 planktotrophic naupliar stages, this second and final stage *viz.* cypris appears. It is enclosed in an oval, bivalved, mussel-shaped shell, with an opening along the ventral margin. The second and third pairs of the appendages of the nauplius disappear in this stage. The first pair of appendages constitutes strong four-jointed antennae, the last segment of which is disc-shaped, and is pierced centrally by a pore, which is the opening of the excretory duct of the cement-glands. The thorax has six pairs of forked natatory limbs on its sides. The abdomen is

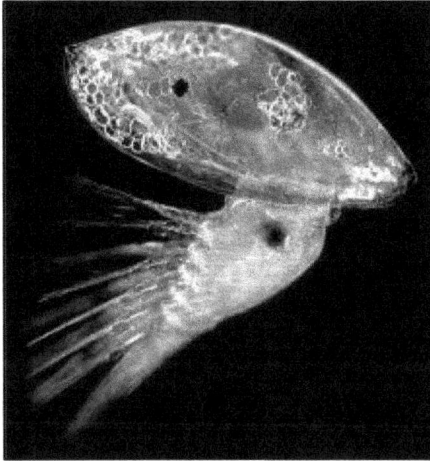

rudimentary, three-jointed, with terminal forked swimming appendages. The cypris larva does not feed, but is nourished by fatty materials stored in the cephalic and dorsal regions of the body. This larva which lasts for a few days seeks suitable surfaces with its modified antennules and attaches using a secreted glycoproteinous substance.

Order: Decapoda

Larvae of Lobsters

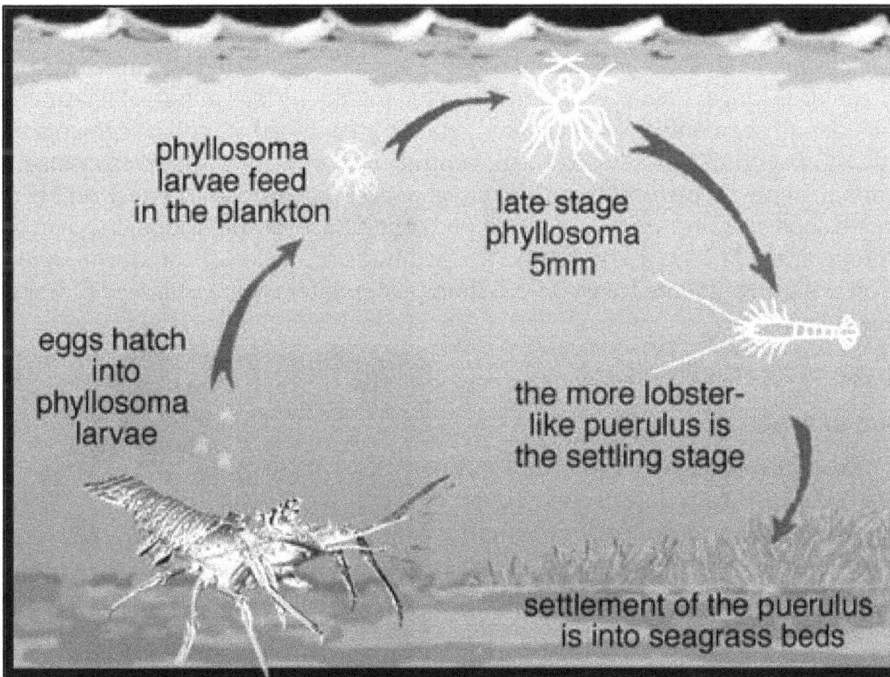

phyllosoma larvae feed in the plankton

late stage phyllosoma 5mm

eggs hatch into phyllosoma larvae

the more lobster-like puerulus is the settling stage

settlement of the puerulus is into seagrass beds

Life Cycle of Lobsters

Phyllosoma of *Thenus orientalis*

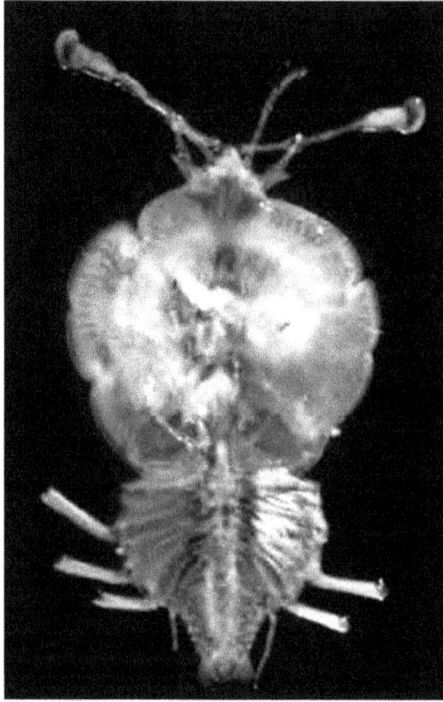

Description: The phyllosoma larva of slipper lobster has a total length of about 1 mm. Its cephalic shield is broader than long. Eyestalk is elongated and unsegmented. Antennule is longer than eyestalk and unsegmented with 3 terminal setae; Antenna is slender, unsegmented with terminal seta. 2nd maxilliped is with 5 segments and without exopod. 3rd maxilliped bears ventral coxal spine and comb-like setae on terminal segment Pereiopods are bigamous with setose exopods. Pereiopod 3 bears short exopod, usually non-setose. Pereiopods 4 and 5 are absent. Abdomen is narrow, approximately 1/2 length of thorax. Uropod buds are absent. Posterior margin of telson is slightly indented at middle. 2 short posterolateral spines flanked by 3 short setae are present.

Larvae of Brachyuran Crab

Zoea of Crab

Description: There are normally 7 zoeal stages and 1 postlarval or megalopal, stage in a typical brachyuran crab's life cycle. The first zoeal stage follows the nauplius stage. It swims using their thoracic appendages *viz.* maxillipeds and pereopods. The zoea larva of a brachyuran crab has a characteristic

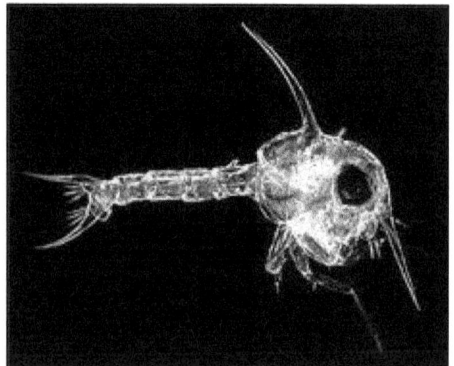

dorsal spine, a rostral spine (the anterior-most spine), and two lateral spines, which all extend from the carapace. These spines help the larva in swimming, and may also be used as a defense against predators. A zoea may be carnivorous, phytoplanktivorous or omnivorous depending on the species. The zoea which measures approximately 0.25 mm at hatching has two stalked compound eyes that are relatively large, compared to the rest of the body and is a filter feeder. Zoeal development may take about 50 days, depending on the existing salinity and temperature. Zoeae moult four to seven times before entering the next stage of development *viz.* megalopa. The final zoeal stage is about 1.0 mm in width.

Megalopa of Crab

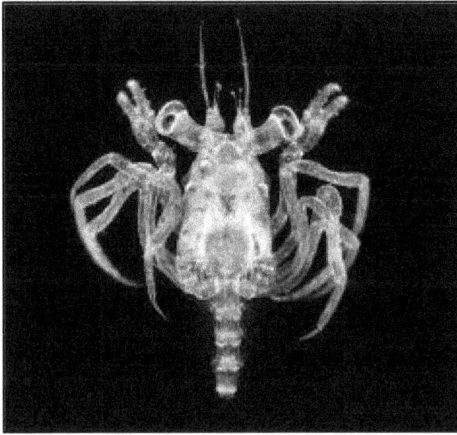

Description: The final moult of the zoeae after about 50 days is characterized by a distinct change to the second larval stage, called a megalopa. The megalops larva is more crab-like in appearance. Its carapace is broader in relation to its length, and has biting claws and pointed joints at the ends of the legs. It measures about 1.0 mm in width. The megalopa swims freely, but generally stays near the bottom in nearshore or lower-estuarine, high-salinity areas. The megalopa stage lasts for about 10 - 20 days, after which it molts into the "first crab" *i.e.* juvenile stage, with the appearance more like that of an adult crab.

Larvae of Hermit Crabs

During the lifecycle of hermit crabs, the zoeae emerged from the eggs become tiny shrimp like creatures and spend their planktonic life in the seas.

The tiny shrimp like creatures undergo about 6 moultings each lasting about a week. Finally, it gets transformed into a creature which resembles a hermit crab and it is known as 'glaucothoe'. It now swims and walks about but remains in the sea where it searches for a shell to live. The baby hermit crab will begin to edge further and further away from the shore where it leads a nocturnal existence. It will undergo

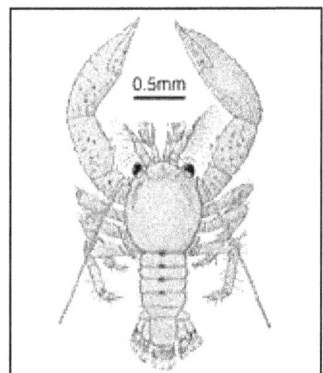

Glaucothoe

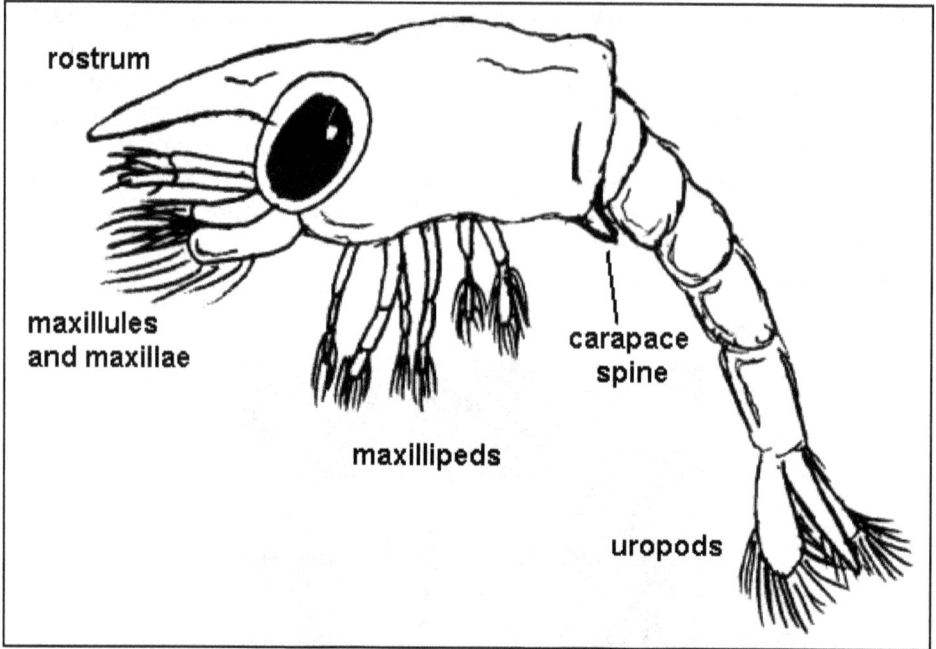

Zoea

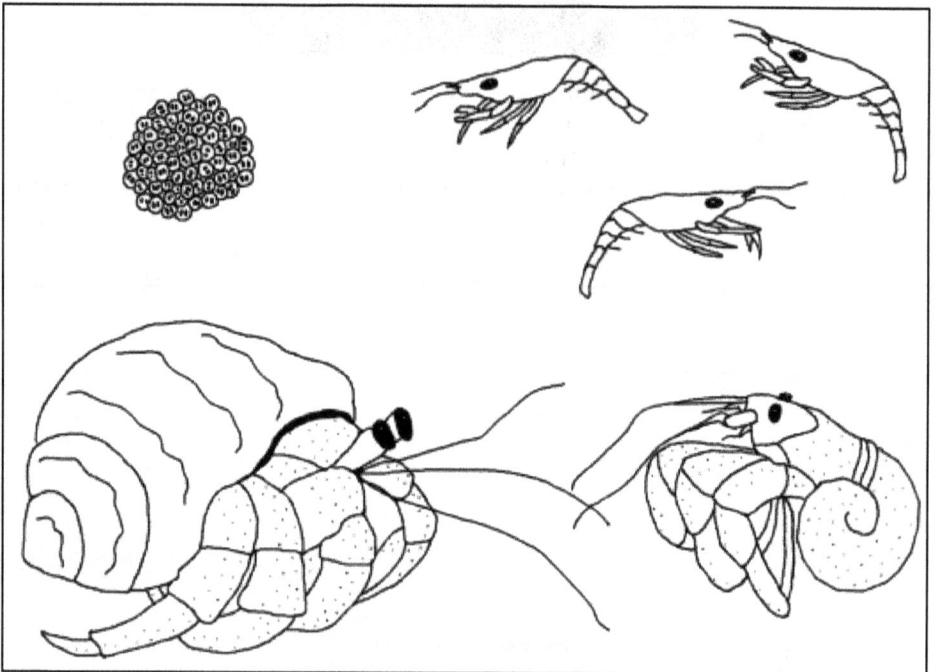

Life Cycle of Hermit Crabs

several moultings, until its body is ready for puberty, which is normally in their second year. It is now ready to breed and so the lifecycle of hermit crabs begins again.

Larvae of Lucifers

Lucifer Mysis

The larval stages of *Lucifer hanseni* include 2 nauliar stages, 3 protozoeal stages, 2 zoeal (mysis) stages and many postlarval stages. The first protozoea is characterised by the presence of 3 small spines at the posterior margin of the carapace and abdominal somites with ventral spines. In protozoea II and III, the eyes are salked. However, the uropod is absent in the protozoea II.

Protozoea III: The total length and carapace length is about 1.0 mm and 0.4 mm respectively. Carapace posterior margin is with 3 small spines. Both rudimentary eyestalks are situated closely together. Third spine of telson is about 3 times as long as 4th one.

12.8. Larvae of Molluscs

Molluscs undergo at least two stages in the plankton before they are metamorphosed to lead a benthic life. The first stage of mollusc development is known as the trochophore larva and is very similar to that of polychaetes. The second stage is the veliger larva where, the velum is a large beautiful winged structure bordered by actively beating cilia. Veligers live in the plankton and grow until their shells become too big and heavy to float. They sink as they undergo a final metamorphosis to a benthic adult.

Trochophores and Veligers of Bivalves

Phylum: Mollusca

Class: Bivalvia

Reproduction in bivalves is normally through the release of gametes into the water column, where the eggs are fertilized and complete larval development takes place. Trochophore larvae first emerge from the eggs. Bivalve veliger larvae developed from the trochophores have flattened bivalve shells, which are more obvious when the velum is retracted.vThough it is difficult to identify species in the early stages, some post-veliger larvae can be identified using their shape and hinge structure.

Size: Trochophores, 50-200 µm, veligers, 65-400 µm.

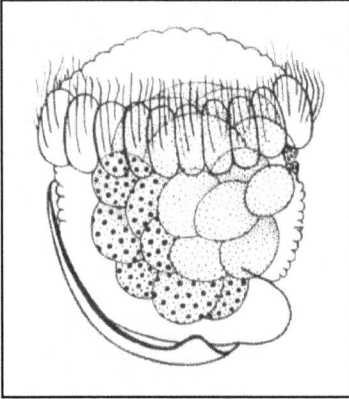 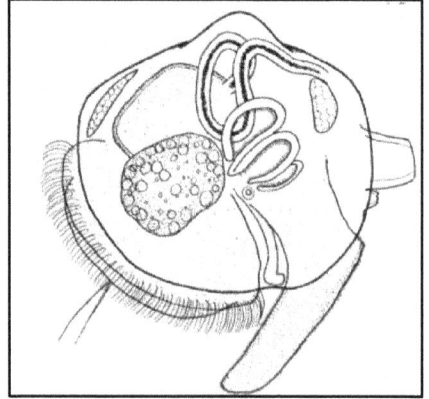

Trochophore   **Veliger**

Gastropod Veligers

Class: Gastropoda

Many gastropod snails like bivalve molluscs pass through a trochophore larval stage before developing into a veliger larva. The gastropod veligers are characterized

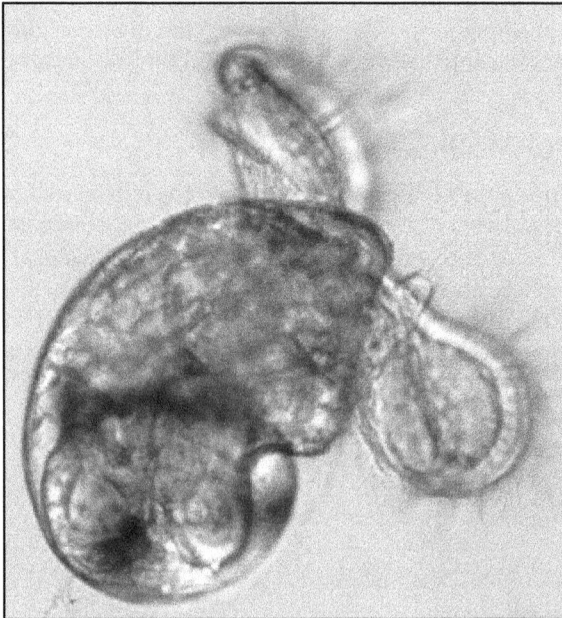

by a shell, foot, and velum which is a lobed, ciliated structure used for swimming and feeding. The velum is derived from the prototroch - a pre-oral ciliated band in the trochophore larva. A dorsal shell gland secretes the shell of the veliger.

Description: The spiralled shell of the veliger in the marine snail, *Nassarius* sp. is with the bi-lobed ciliated velum extended from the shell. A secondary (post-oral) ciliated band known as the metatroch is also seen below the main ciliated band of the velum.

Paralarvae of Cephalopods

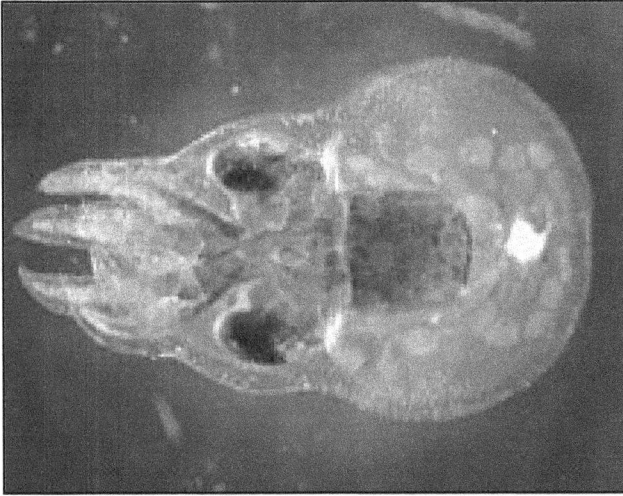

Paralarva of *Loligo* sp.

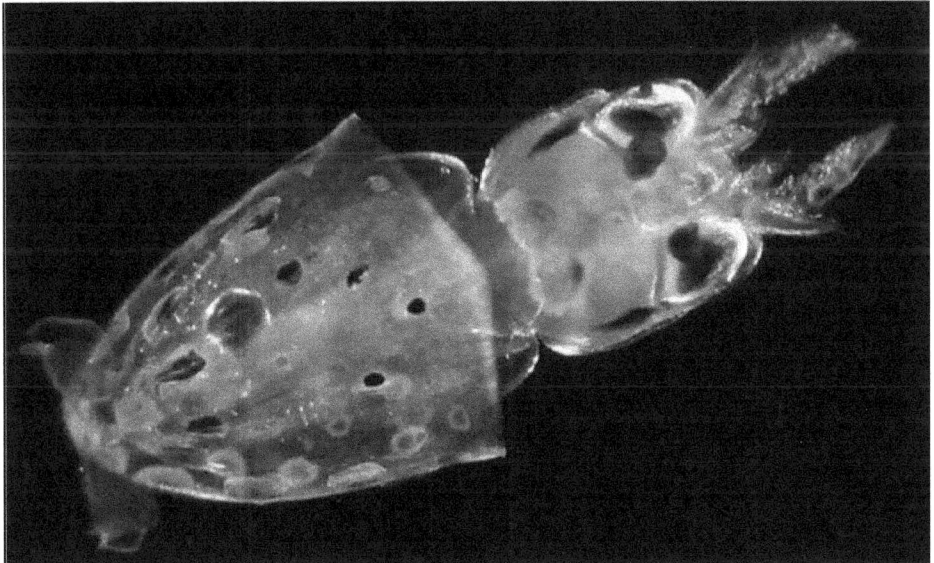

Advanced Paralarva of *Loligo* sp.

Paralarvae of *Octopus* sp.

Description: Recently hatched *Loligo* sp. measured about 2.3 mm. Over a period of 45 days, the paralarvae (the stage between the hatchling and subadult) increased in length up to 18.5 mm. Most of the paralarvae aggregate near the water surface towards the light, and swim constantly. Paralarvae feed on mixed zooplankton and live up to 45 d, reaching mantle length of 14 mm.

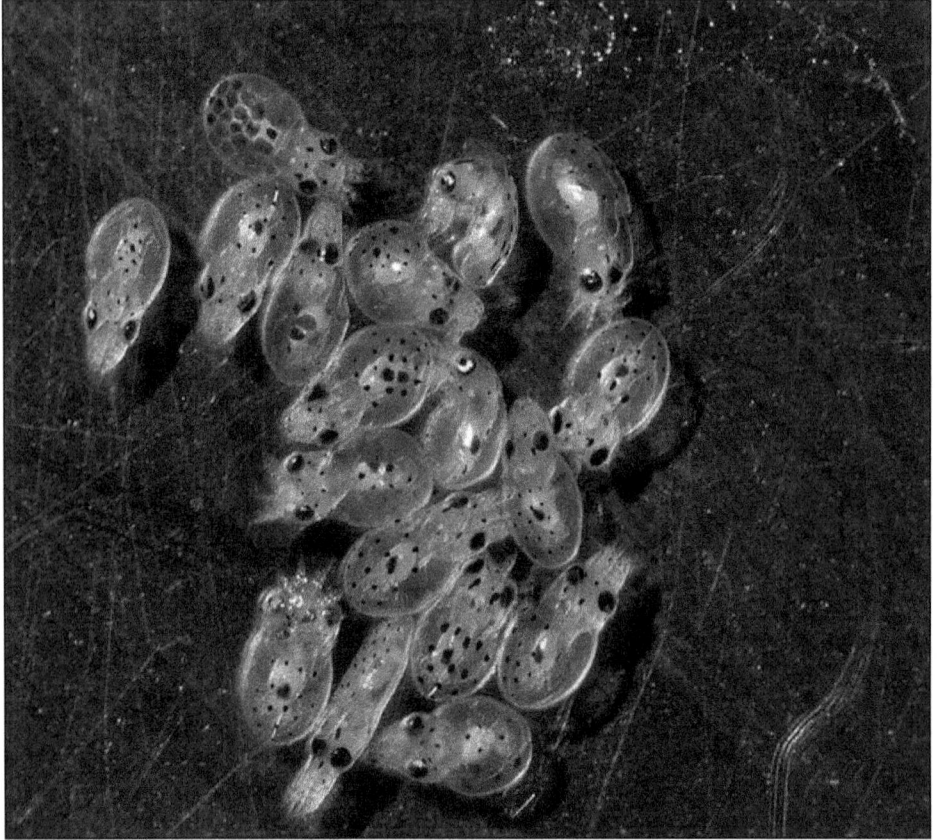

Hatchlings of *Octopus* sp.

Paralarva of *Octopus* sp.

Description: Newly hatched young of the benthic, coastal-living octopod, *Octopus* sp. enter the plankton and remain there for perhaps 8 weeks. At hatching, the arms are short and possess a few, large, primary suckers. The buccal mass is relatively large in proportion to the size of the animal. The eyes are large. The central nervous system has fairly well-defined lobes, some of which develop earlier than others.

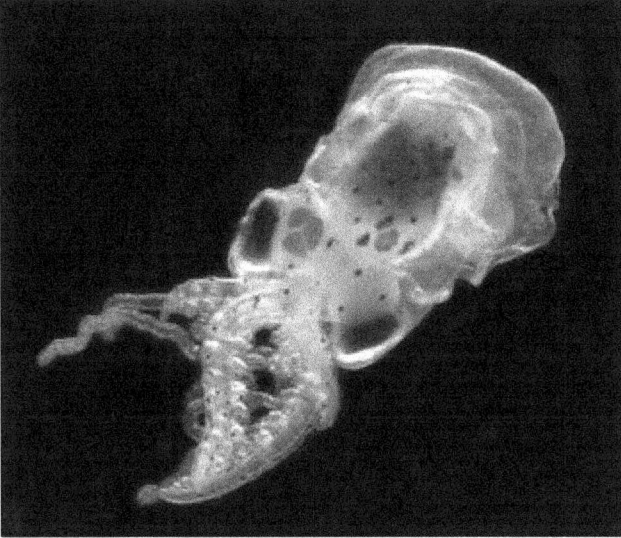

Paralarva of *Octopus* sp.

12.9. Larvae of Echinoderms

Echinoderms are entirely benthic animals and are among the dominant phyla of the abyssal benthos. They are also very abundant on the seafloor in nearshore waters. In order to reproduce themselves and distribute the species, all five groups of Echinoderms use planktonic larvae.

Bipinnaria and Brachiolaria Larvae of Starfish

Phylum: Echinodermata

Class: Asteroidea

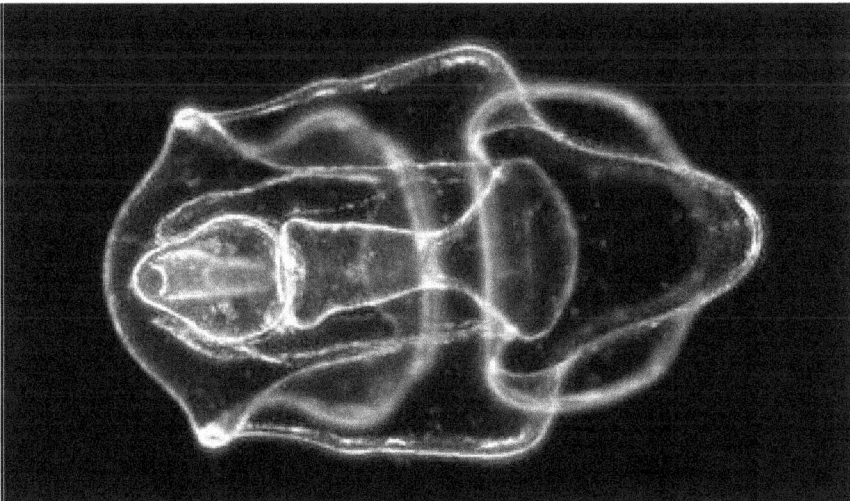

Bipinnaria Larva

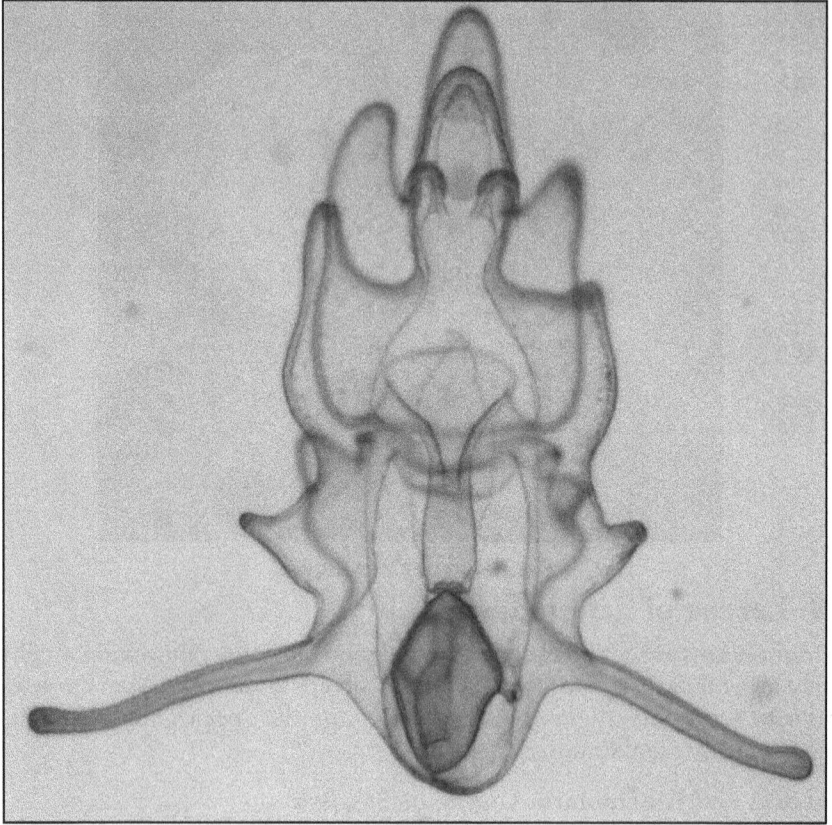

Brachiolaria Larva

Description: The starfishes which are with small, non-yolky eggs have a free-swimming, feeding, bipinnaria larva. The latter has two unconnected ciliated bands, extending over numerous lobes that later develop into hollow arms unsupported by skeletal rods. One of the ciliary bands is much longer than the other and loops over most of the body, while the other is short and restricted to a small area of the body. The bipinnaria larva superficially resembles the auricularia larva of holothurians but the ciliary band in the auricularia is continuous and not in two parts. Some species metamorphose directly to a juvenile from the bipinnaria stage, while other bipinnaria become a brachiolaria larva which is a further developmental stage rather than a separate larval type.

The brachiolaria has a single dorsal and two ventral stubby brachiolar arms, the only feature that distinguishes the brachiolaria from bipinnaria larvae. The brachiolaria produces adhesive secretions used in temporary attachment to the substrate. There is also an adhesive disc situated between the arms for more substantial attachment. The brachiolar arms are less obvious among the other arms in later larvae. Some starfishes with large yolky eggs have a non-feeding, pelagic brachiolaria stage. These morphologically simpler larvae lack ciliary bands and arms. In both bipinnaria and brachiolaria larvae, metamorphosis starts in the plankton and an obvious small juvenile starfish develops from the larva. The arms of the juveniles are likely to be more rounded and less projecting. Some of the longer arms begin to shorten as they are reabsorbed. Most brachiolaria larvae need to settle and attach to the substrate before metamorphosis can continue. The benthic juvenile then detaches from the larval body.

Ophiopluteus Larva of Brittle-Star

Class: Ophiuroidea

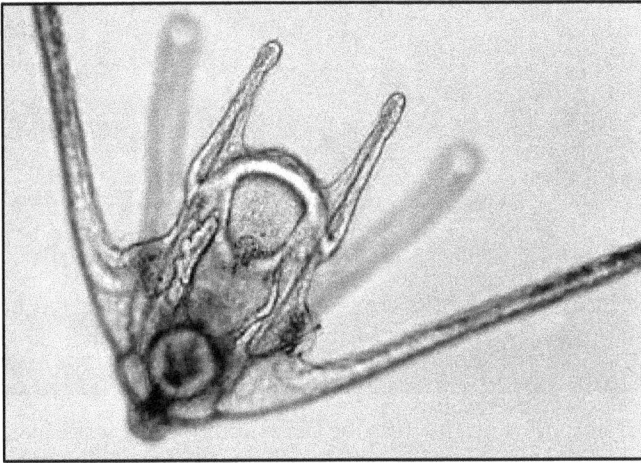

Description: Ophiopluteus of brittle star has 4 pairs (occasionally 5 or 6) of arms, supported by delicate, calcareous skeletal rods. A continuous band of cilia follows the contours of the arms, used in locomotion and feeding. The larvae swim with the arms pointed forwards. Some have very long thin arms and compact bodies while others have fleshy squat arms. Red pigment cells are often found scattered over the body.

The initial larval body is triangular and the skeletal rods may be seen internally. The two posterolateral arms develop first and are longer than the other arms and extend laterally. Each half of the larval skeleton comprises rods supporting the arms. In some advanced larvae, ciliated lobes used in locomotion called epaulettes are located at the junction between the posterolateral and post-oral arms.

Echinopluteus Larva of Sea-urchin

Class: Echinoidea

Description: The Echinoplutei larvae of sea urchins superficially resemble the ophioplutei larvae of brittle stars. However, the arms are not strictly equivalent in the two groups. Both groups generally have 4 pairs of arms, but most echinoplutei do not have arms corresponding to the pair of longer posterolateral arms of ophioplutei. The longest arms in advanced echinoplutei are the pairs of post-orals and posteriodorsals

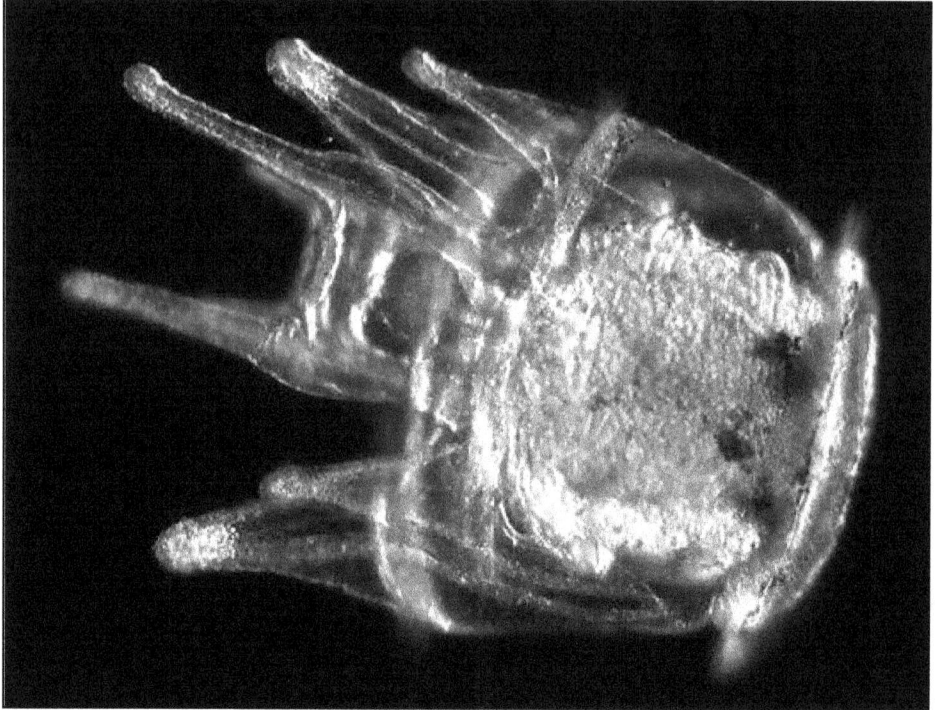

that extend dorsally and ventrally from the larval anterior/posterior axis. The skeleton in echinoplutei is usually more complex than in ophioplutei. Further, the bodies of echinoplutei are generally laterally flattened, while ophioplutei bodies are more dorso-ventrally flattened.

Auricularia, Doliolaria and Pentacula Larvae of Sea Cucumbers

Class: Holothuroidea

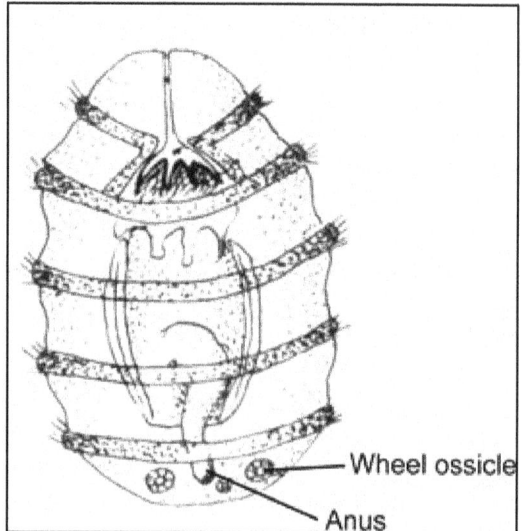

**Doliolaria Larva of
Sea Cucumber**

Wheel ossicle

Anus

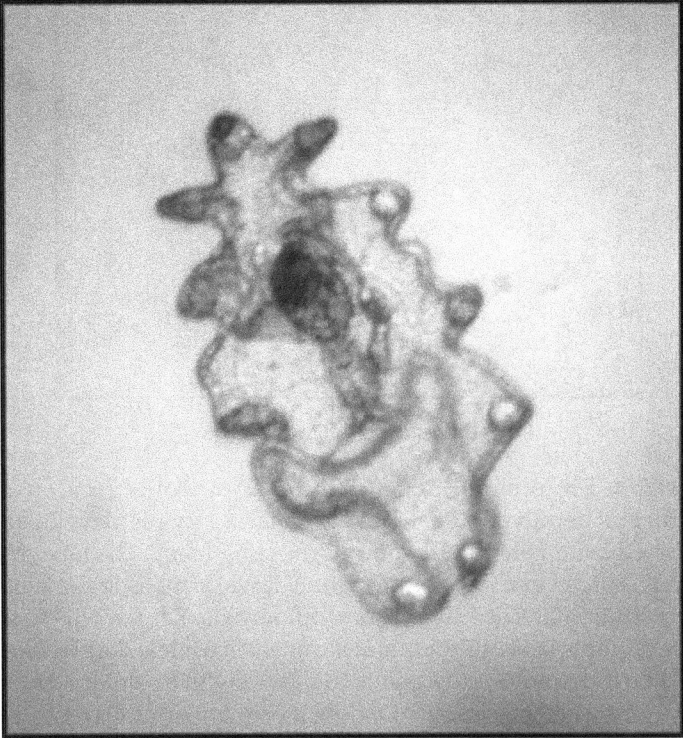

Auricularia Larva of Sea Cucumber

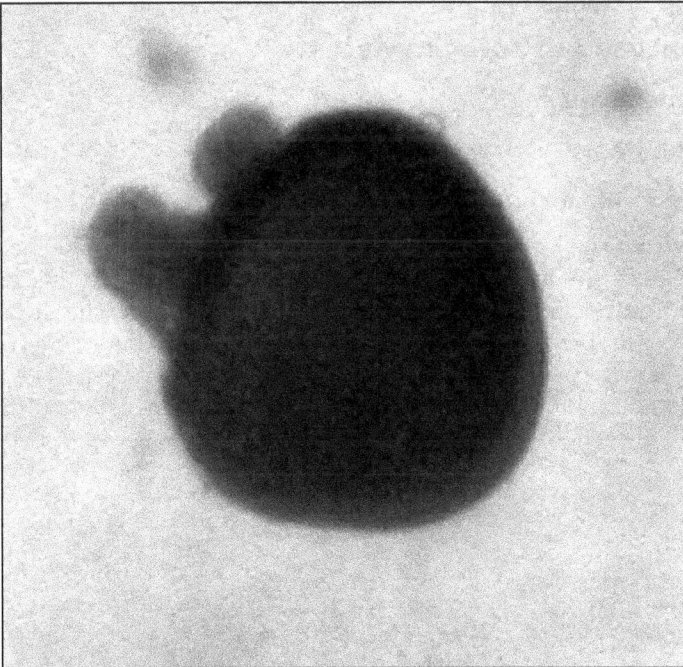

Pentacula Larva of Sea Cucumber

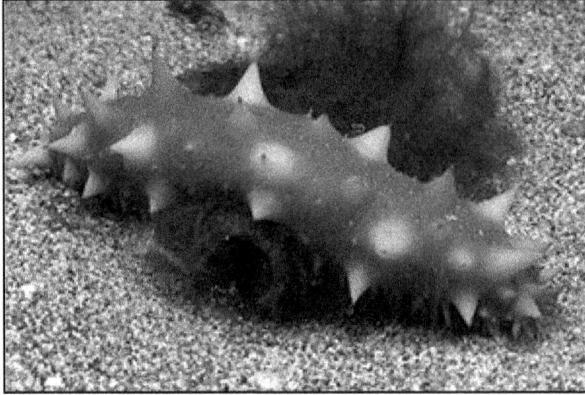

Juvenile Sea Cucumber

Description: From small eggs of the holothuroids (sea cucumbers), a planktotrophic, bilaterally symmetrical auricularia larva is developed. It has an elaborate ciliated band extending around projecting lobes. The lobes may become very numerous, but never develop into distinct larval arms as in the later bipinnaria larvae of asteroids or plutei of echinoids and ophiuroids. The auricularia superficially resembles the bipinnaria of asteroids and it transforms into a non-feeding doliolaria larva which then transforms into a pentactula larva with 5 primary tentacles (which develop into the ring of tentacles surrounding the mouth in adults) and 1 or 2 primary feet. This is the settlement stage, but it may remain in the plankton for some time before metamorphosing into a juvenile.

12.10. Larvae of Protochordates

Larva of *Amphioxus* sp.

Phylum: Chordata

Subphylum: Cephalochordata

Description: The larvae of *Amphioxus* are small and ciliate like the planktonic larvae of marine invertebrates. The characteristic feature of this larva is the development of functional muscles at the early larval stage. The duration of the metamorphosis in *Amphioxus* is not so short unlike other marine invertebrates. Rapid changes found in the mouth and branchial systems seem to correspond with intrinsic requirements. Locomotion and feeding behaviour do not change substantially between

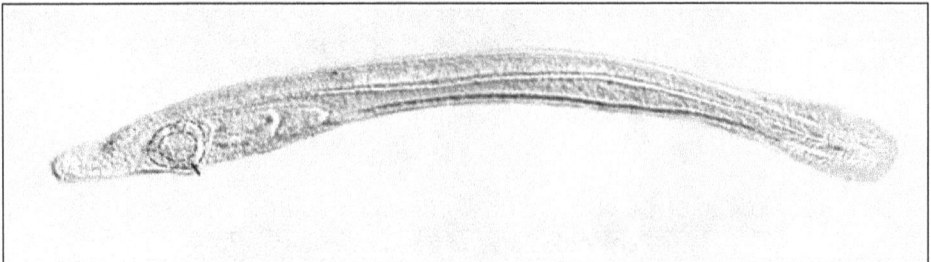

pre- and post metamorphosis in lancelets. The European lancelet, *Branchiostoma lanceolatum* stops its growth or reduces its the body length during metamorphosis. On the other hand, the Florida lancelet, *Branchiostoma floridae* continues its growth.

Subphylum: Hemicordata

Tornaria Larva of *Balanoglossus* sp.

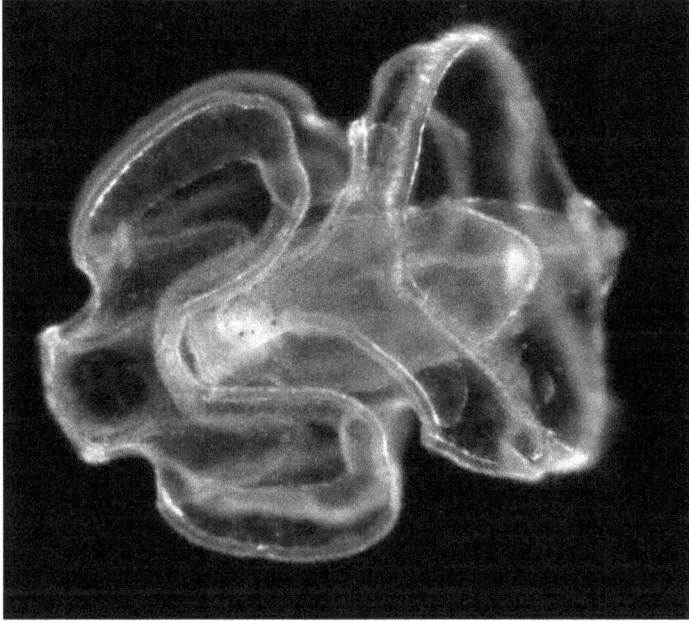

Description: Tornaria larva of *Balanoglossus* measures about 3 mm. It is glossy in appearance with an oval body and ventral mouth. The cilia form two bands on the body surface. The anterior ciliary band or circumoral band has a winding course over the preoral surface and forms a postoral loop, the cilia of which are short. The posterior ciliated band or telotroch occurs as a ring in front of the anus and its cilia are long and serve as locomotor organs. At the anterior end is an apical plate of thickened epidermal cells, which bears a pair of eye spots or ocelli and a tuft of sensory cilia called apical tuft or ciliary organ. The protocoel (proboscis coelom) is in the form of a thin-walled sac and opens to the exterior through a hydropore. To the right of the hydropore lies a pulsating heart vesicle. The collar and trunk coeloms appear only in the older larva.

Subphylum: Urochordata

Tadpole Larva of Tunicates

Description: The tadpole larvae of tunicates have adhesive papillae at the front which secrete a sticky attachment substance. Once attached, metamorphosis is rapid. Within minutes, the tail gets reabsorbed and the larva rotates to bring the siphonal openings into correct orientation, and the branchial-basket filtering system becomes

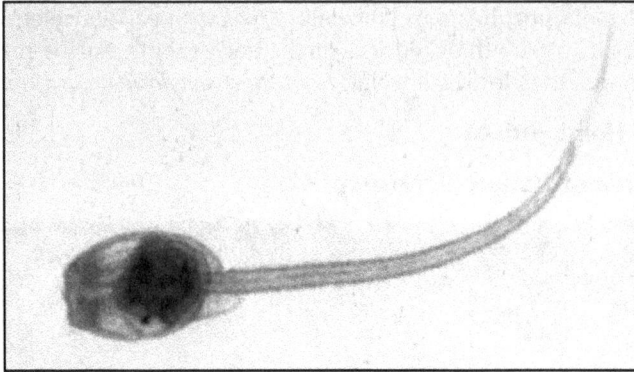

functional. In addition to the tail being reabsorbed, other structures useful in swimming and navigation in the larva are resorbed or reduced in size. The structures such as notochord, dorsal tubular nerve cord, tail, adhesive papillae, and various ganglia and sensory organs (*e.g.*, ocellus or eyespot) help in swimming, sensory input, and attachment, and are lost or resorbed during the metamorphosis.

12.11. Eggs and Larvae of Fishes (Ichthyoplankton)

Phylum: Chordata

Class: Pisces

The most conspicuous and abundant members of the Phylum Chordata are the modern, bony fishes (subphylum: Vertebrata, Class: Osteichthys). Most families of fish cast their eggs into the plankton where they hatch into larvae. Ichthyoplankton are the eggs and larvae of fish. They are usually found in the epipelagic or photic zone. Ichthyoplankton are planktonic, meaning they cannot swim effectively under their own power, but must drift with the ocean currents. Fish eggs cannot swim at all, and are unambiguously planktonic. Early stage larvae swim poorly, but later stage larvae swim better and cease to be planktonic as they grow into juveniles. Fish larvae

Description of Different Stages of Ichthyoplankton

Egg stage	Spawning to hatching
Larval stage	From hatching till all fin rays are present and the growth of fish scales has started.
	The larval stage may be further subdivided into pre-flexion, flexion, and post-flexion stages. In many species of fishes, the body shape and fin rays, as well as the ability to move and feed, develops most rapidly during the flexion stage.
Juvenile stage	This stage starts with all the fin rays being present and scale growth underway, and completes when the juvenile becomes sexually mature or starts interacting with other adults.
Yolk-sac larval stage	From hatching to absorption of the yolk-sac
Transformation stage	From larva to juvenile. This metamorphosis is complete when the larva develops the features of a juvenile fish.

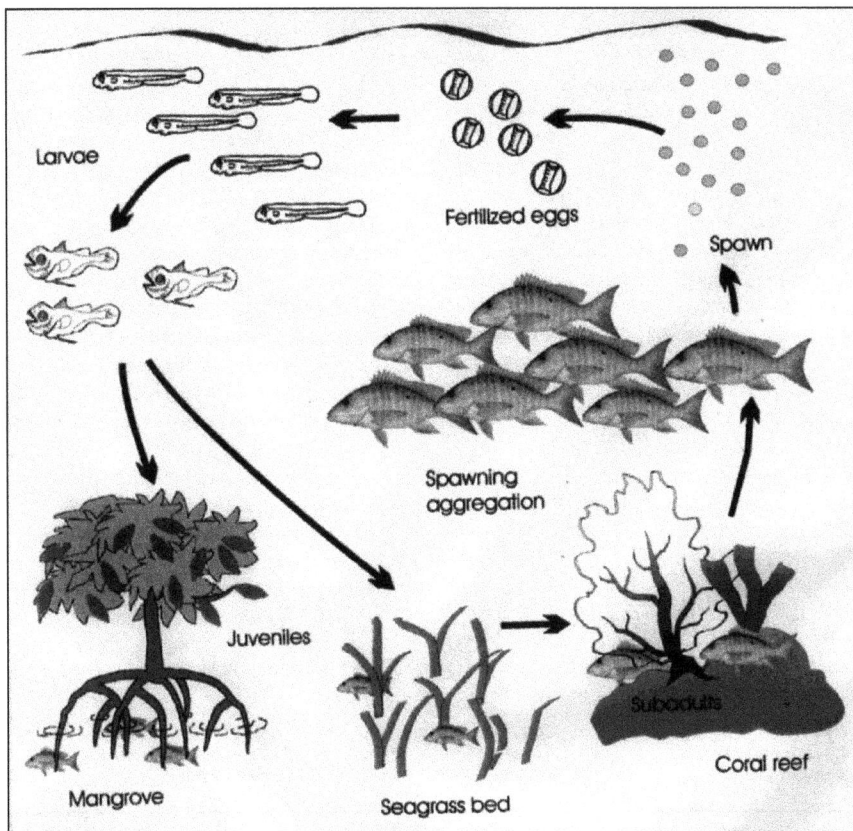

Life Cycle of Fishes.

Larvae

Fertilized eggs

Spawn

Spawning aggregation

Juveniles

Mangrove

Seagrass bed

Subadults

Coral reef

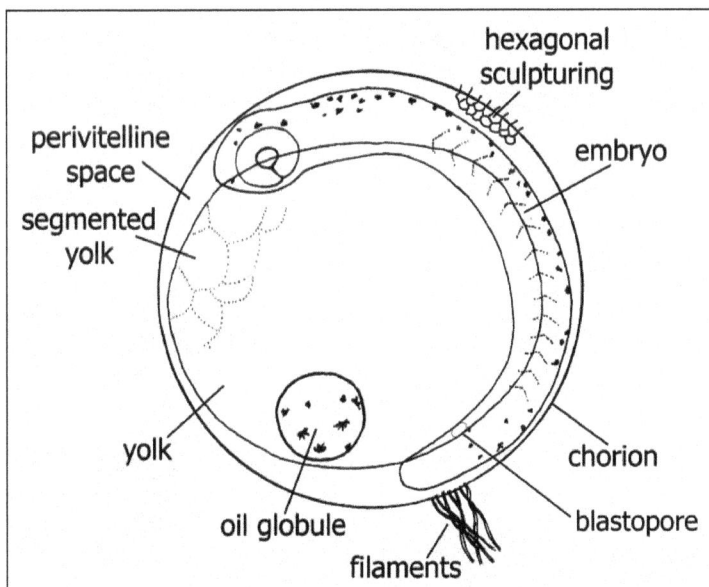

Structure of a Fertilized Fish Egg.

hexagonal sculpturing

perivitelline space

embryo

segmented yolk

yolk

chorion

oil globule

blastopore

filaments

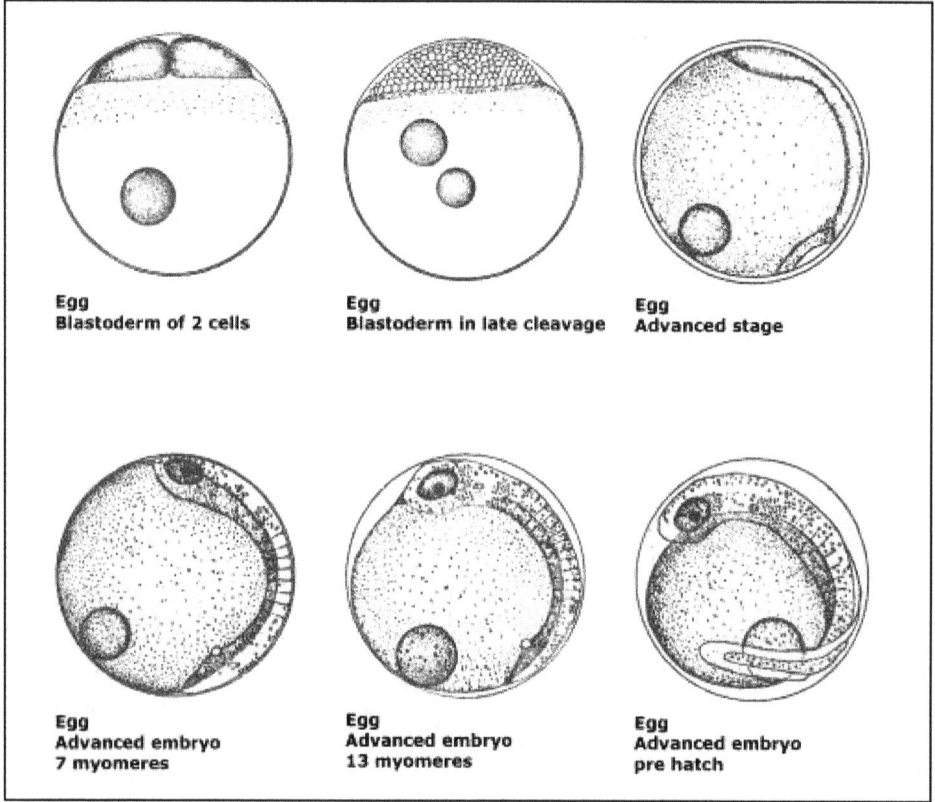

Egg
Blastoderm of 2 cells

Egg
Blastoderm in late cleavage

Egg
Advanced stage

Egg
Advanced embryo
7 myomeres

Egg
Advanced embryo
13 myomeres

Egg
Advanced embryo
pre hatch

Fertilization in a Fish Egg.

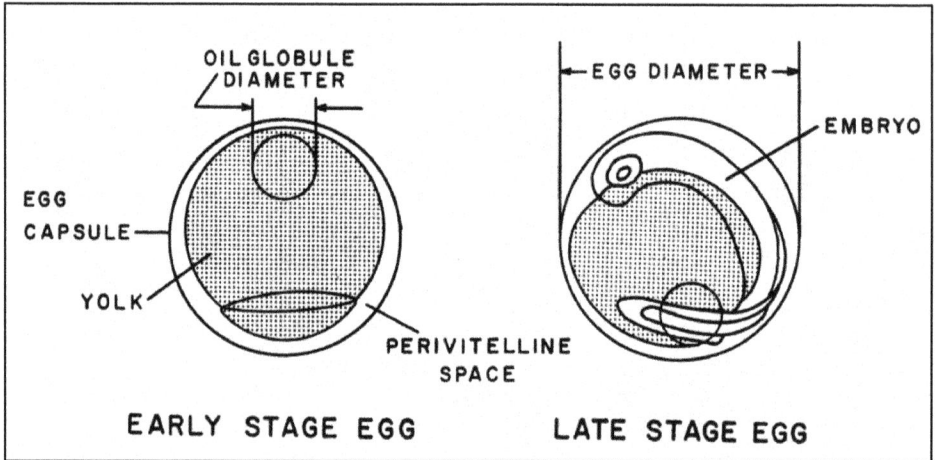

OIL GLOBULE
DIAMETER

EGG DIAMETER

EMBRYO

EGG
CAPSULE

YOLK

PERIVITELLINE
SPACE

EARLY STAGE EGG

LATE STAGE EGG

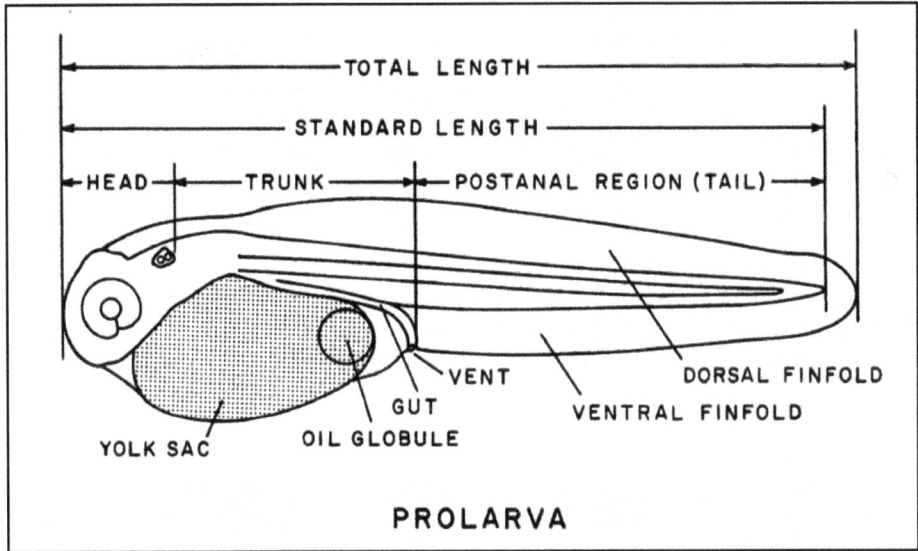

TOTAL LENGTH

STANDARD LENGTH

HEAD — TRUNK — POSTANAL REGION (TAIL)

VENT
GUT
YOLK SAC   OIL GLOBULE
DORSAL FINFOLD
VENTRAL FINFOLD

PROLARVA

Eggs and Larval Stages of *Sardinella* sp.

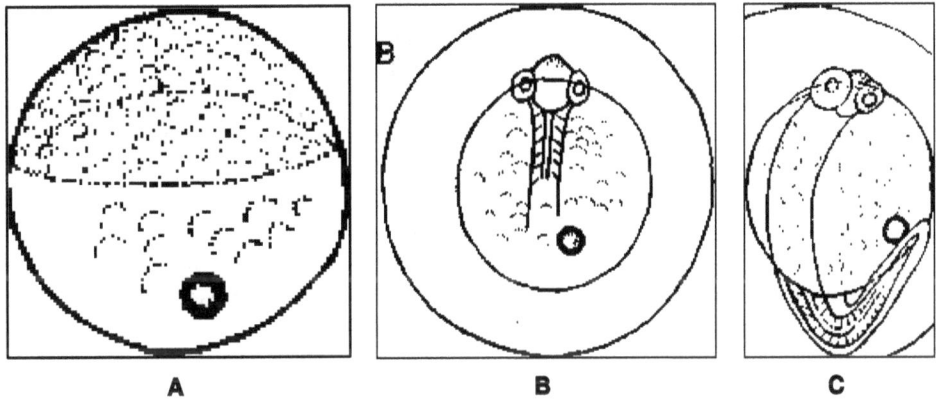

A   B   C

A. Ealy Egg; B. Middle Egg; C. Late Egg.

are part of the zooplankton that eats smaller plankton, while the fish eggs carry their own food supply. Both eggs and larvae are themselves eaten by larger animals.

Description: The eggs of this species are spherical and are 1.3 - 1.6 mm in diameter. In living condition, the size of the transparent, vacuolated and spherical yolk ranges from 0.8 to 0.9 mm and the single, golden-yellow oil globule from 0.1 to 0.13 mm in diameter. In the embryonic development, three stages are discernible; the early egg (Figure A) with blastoderm, the middle egg (Figure B) with a well defined anterior half of the body and the late egg (Figure C) with fully developed embryo. Pigmentation on the embryo is absent in the early egg, appeared as a few minute black spots on the

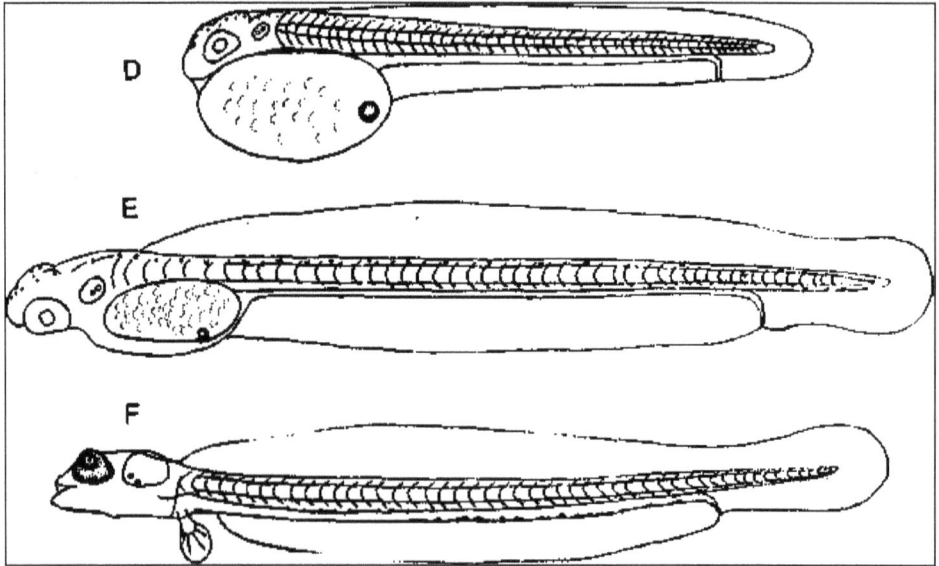

D. Newly hatched Larva; E. 18' hrs Old Larva; F. Early Postlarva (41 hrs old).

dorsal side of the embryo in the middle egg and increased in number in the late egg. The newly hatched larvae (Figure D) has a total length ranging from 2.9 to 3.2 mm. The globular yolk is rounding off posteriorly and the oil globule is situated towards its hinder end. Minute black pigments are seen on the dorsal side of the anterior half of the body. There are 40 preanal and 8 postanal myomeres. The 18 hours old larva (Figure E) has a total length ranging from 4.4 to 4.5 mm. The mouth is not yet formed. The yolk is smaller and the dorsal pigment spots are more prominent, with a few appearing on the posterior half of the larva. A few pigments appear ventrally in the postanal region also. The number and disposition of the myomeres remain the same. Post-larva (Figure F) of this species has a size range of about 20mm and juveniles 30 mm.

Literature Cited

Anon., 1968. *Zooplankton Sampling*. UNESCO Monogr. Oceanogr. Methodology, 2.

Fraser,J. 1962. *Nature Adrift, The Story of Marine Plankton*. G.T. Fowlis and Co., London.

Hardy, A.1956. *The Open Sea: Its Natural History*.Vol.1. The World of Marine Plankton. Houghton Mifflin, Boston.

Santhanam, R. and A. Srinivasan, 1994. *A Manual of Marine Zooplankton*. Oxford and IBH publishing Co. Pvt. Ltd., New Delhi.

Wickstead, J.H., 1965. *An Introduction to the Study of Tropical Plankton*.Hutchinson and Co., London.

Wimpenny, R.S., 1966. *The Plankton of the Sea*. Elsevier, New York.

Index

A

Acanthometron sp. 104

Acartia (Acanthacartia) *fossae* 157

Acartia erythraea 157

Acid cleaning 21

Acidified formaldehyde 17

Acnidaria 130

Acnidarians 130

Acrocalanus longicornis 146

Actinotrocha larva 182

Adult stages 185

Aequorea pensilis 123

Akashiwo sanguinea (Hirasaka) Hansen and Moestrup 77

Alexandrium catenella 89

Amnesic Shellfish Poisoning (ASP) 92

Amphidinium sp. 100

Amphioxus sp. 204

Amphipoda 142

Amphipods 142

Amphorellopsis acuta 119

Amphorides amphora 120

Animalia 2

Annelida 134

Appendicularia 173

Arrow worms 136

Arthropoda 138

Asterionellopsis 46

Asterionellopsis glacialis (Castracane) Round 66

Asteroidea 199

Asterompalus flabellatus (Brebisson) Greville 50

Asterompalus wyvillei Castracane 50

Atentaculata 131

Aurelia aurita 128, 178

Azaspiracid Poisoning (AZP) 91

B

Bacillaria paxillifera (O. F. Müller) Hendey 70

Bacillariophyceae 1, 49

Bacteriastrum delicatulum Cleve 62

Balanoglossus sp. 205

Barnacles 190

Bellerochea malleus (Brightwell) Van Heurck 57

Beroe sp. 131

Beroidea 131

Biddulphia 48

Biddulphia mobiliensis Bailey 54

Bilateral symmetry 47

Binary fission 48

Bivalvia 195

Bottling 20

Brachinus calyciflorus 132

Brachionus plicatilis 133

Brachiopods 181

Brachyuran crab 192

Bryozoans 182

C

Calanoida 146, 184

Calanopia elliptica 153

Calibration 41

Canada balsam 22

Cavolinia tridentata 169

Centropages furcatus 147

Centropages indicus 149

Cephalochordata 204

Cephalopods 197

Cerataulina pelagica (Cleve) Hendey 55

Ceratium breve (Ostenfeld and Schmidt) Schroder 80

Ceratium contortum (Gourret) Cleve 80

Ceratium furca (Ehrenberg 1836) Claparéde and Lach 77

Ceratium fusus (Ehrenberg 1834) Dujardin 1841 78

Ceratium karstenii Pavillard 80

Ceratium pulchellum B. Schroder 79

Ceratium trichoceros (Ehrenberg) Kofoid 1908 78

Ceratium tripos (O. F. Müller 1781) Nitzsch 1817 79

Chaetoceros 46

Chaetoceros affinis Lauder 62

Chaetoceros curvisetum Cleve 63

Chaetoceros didymus Ehrenberg 64

Chaetoceros diversus Cleve 64

Chaetoceros lorenzianus Grunow 63

Chaetoceros spp. 21

Chaetognatha 3, 136

Chemical method 25

Chloromonads 1

Chlorophyceae 2, 86

Chlorophyll *a* 32

Chordata 204, 206

Chrysoara colorata 128

Ciguatoxin 94

Circular net 30

Cirripedia 190

Cladocerans 1, 136

Clarke-Bumpus Horizontal Closing Net 13

Clearing 19

Climacodium frauenfeldianum Grunow 56

Clione limacina 171

Closing nets 13

Clytemnestra scutellata 166

Cnidaria 123

Cnidarians 122, 178

CO_2 4, 37

Cochlodinium polykrikoides 97

Codonellopsis ostenfeldi 113

Coelenterata 122, 178

Coenogastropoda 168

Collection 5

Copepod feeding 143

Copepoda 145

Copepodid 185

Copilia mirabilis 161

Corethron 48

Corethron criophilum Castracane 58

Corycaeus dahli 161

Coscinodiscus 48

Coscinodiscus eccentricus 49

Coscinodiscus jonesianus (Greville) Ostenfeld 52

Coscinodiscus oculus-iridis Ehrenberg 53

Coscinodiscus radiatus 46

Counting 26

Counting chambers 28

Coxliella annulata 113

Creseis acicula 168

Crustacea 138, 184

Cryptophyceae 1, 84

Cyanea capillata 129

Cyanophyceae 1, 87

Cyclomyaria 177

Cyclopoida 160

Cyclotella 48

Cyclotella meneghiniana Kutzing 49

Cyclotella striata (Kutzing) Grunow 49

Cydippida 130

Cyphonautes Larva 183

Cypridina acuminata 141

Cypridina dentata 141

Cypridina meditteranea 140

Cypris 190

D

Dadayella cuspis 120

Decapoda 167, 191

Desmomyaria 176

Diarrhetic Shellfish Poisoning 88

Diatoms 46

Dictyocha speculum Ehrenberg 1839 85

Dictyochophyceae 1, 85

Dictyocysta seshaiyai 112

Digital flow meter 30

Dinoflagellates 72

Dinophyceae 1, 46

Dinophysis caudata Saville-Kent 1881 75

Dinophysis miles Cleve 76

Diphyes dispar 124

Diploneis weissflogii (A. Schidt) Cleve 69

Displacement volume 23

Dissection 22

Ditylum brightwelli (West) Grunow 57

Dolioletta gegenbauri 177

Doliolids 177

DPX mount 22

Dunaliella salina (Dunal) Teodoresco 1905 86

E

Echinodermata 199

Echinoderms 199

Echinoidea 201

Echinopluteus larva 201

Ectoprocta 183

Eirene viridula 123

Ethanol 22

Eucalanus attenuatus 159

Eucampia zodiacus Ehrenberg 56

Euchaeta rimana 147

Euglenophyceae 2, 86

Euterpina acutifrons 166

Eutintinnus conicus 121

Eutintinnus tenue 121

Exuviaella compressa Barley and Ostenfeld 74

F

Faunal enumeration 25

Favella brevis 117

Favella campanula 118

Favella ehrenbergi 118

Folsom plankton splitter 25

Foraminifera 103

Formaldehyde 19
Fragilariopsis 46
Friedinger's water sampler 7
Frittilaria borealis 174

G

Gastropoda 196
Glaucus atlanticus 172
Globoquadrina pseudofoliata 103
Globorotalia pseudobulloides 104
Glutaraldehyde solution 18
Gonyaulacales 83
Gonyaulax polygramma 97
Gravimetric methods 24
Guinardia flaccida (Castracane) Peragallo 61
Gymnodinium catenatum (Dinophyceae) 89
Gymnosomata 171
Gyrosigma balticum (Ehrenberg) Rabenhorst 68

H

Hardy's Continuous Plankton Recorder 14
Harpactocoida 164
Helicostomella longa 114
Hemiaulus sinensis Greville 56
Hemicordata 205
Hemidiscus cuneiformis Wallich 53
Hermit crabs 193
Heterosigma akashiwo 100
Heterosigma sp. 84
Holoplankton 102
Holothuroidea 202
Homo-yessotoxin 93
Hydrozoa 123
Hyperia galba 143
Hyperia macrocephala 142
Hyrax 22

I

Ichthyoplankton 206
Ichthyotoxic Species 95
Importance 3
Indian Ocean standard net 12
Inverted-microscope method 26

J

Janthina janthina 168

K

Karenia brevis (Davis) Hansen and Moestrup 2000 77
Karenia mikimotoi 99

L

Labelling 20
Labidocera acuta 153
Larvacea 173
Larvaceans 173
Lauderia annulata Cleve 51
Leptocylindrus danicus Cleve 64
Leptocylindrus minimus Gran 65
Leptomedusae 123
Light and dark bottle method 34
Light microscopy 21
Limacina 170
Limacina bulimoides 171
Limacina helicina 170
Lingula larva 181
Liriope tetraphylla 124
Lobsters 191
Longipedia coronata 164
Lucifer hanseni 167
Lucifer mysis 195
Lugol's solution 18
Luminella sp. 115

M

Macrosetella gracilis 165
Magnification 39

Margalef's diversity index 43

Marine rotifers 131

Marine zooplankton 3

Marine zooplankton biodiversity 4

Megalopa 193

Megaplankton 1

Melosira 48

Meroplankton 178

Mesoplankton 1

Metacylis jorgensenii 115

Metacylis tropica 116

Meyer's water sampler 6

Micrometry 39

Microplankton 1, 9

Microsetella norvegica 164

Mollusca 195

Monogononta 131

Mounting 22

Mounting techniques 21

Müller's larva 179

Mysis stages 189

N

Nanoplankton 1, 5

Nansen reversing water sampler 7

Narcomedusae 124

Narcotisation 18

Naupliar stages 184, 186

Nauplius 190

Navicula directa Cleve 69

Nemertine worms 180

Neutralized formaldehyde 17

Nitzschia closterium (Ehrenberg) W. Smith 71

Nitzschia longissima (Brébisson) Ralfs in Pritchar 70

Nitzschia sigma (Kutzing) W. Smith 71

Noctiluca scintillans 99

Nudibranchia 172

O

Ocean biogeochemical 4

Octopus sp. 198

Ocular micrometer 39

Odontella sinensis (Greville) Grunow 54

Oikopleura dioica 173

Oikopleura longicauda 174

Oithona attenuata 160

Oithona plumifera 160

Oncaea clevei 161

Opephora schwartzii (Grunow) Petit 66

Ophiuroidea 201

Ordinary flow meter 30

Ornithocercus magnificus Stein 1883 76

Osmic acid 18

Ostracoda 140

Ostreopsis siamensis 95

P

Paracalanus indicus 146

Paralytic shellfish poisoning 88

Pectenotoxin (PTX) 93

Pelagobia longicirrata 134

Penaeid prawn 186

Penilia avirostris 138

Phalacroma argus Stein 74

Phalacroma cuneus Schutt 74

Phoronids 182

Phyllodocida 134

Physalia physalis 126

Phytoplankton 3, 17, 21, 46

Pilidium larva 180

Pisces 206

Plagioselmis sp. 84

Plankton 1, 34

Plankton processing 17

Plankton pumps 8

Planktonic annelids 133

Planktonic flatworms 179
Planktonic foraminiferans 103
Planktoniella sol (Wallich) Schütt 50
Plant carotenoids 33
Platyhelminthes 179
Pleurobrachia pileus 130
Pleurosigma elongatum Smith 69
Pleurosigma galapagense Cleve 69
Ploimida 131
Plotohelmis capitata 134
Podolampas palmipes Stein 1883 82
Podon polyphemoides 138
Polychaeta 134
Pontella danae var. *ceylonica* 155
Pontellopsis herdmani 155
Porpita porpita 126
Prasinophyceae 2, 86
Pre-lab oreparation 34
Preservation 17, 19
Proboscia alata (Brightwell) Sundström 60
Prorocentrum micans Ehrenberg 1833 73
Prorocentrum rostratum Stein 73
Protoceratium reticulatum 93
Protochordates 173, 204
Protoperidinium crassipes 91
Protoperidinium depressum (Bailey 1855) Balech 81
Protoperidinium murrayi Kofoid 82
Protoperidinium ovatum (Pouchet) Schutt 82
Protozoa 2
Protozoans 102
Protozoeal stages 187
Prymnesiophyceae 2, 85
Prymnesium parvum (Golden algae) 98
Pseudodiaptomus arabicus 150
Pseudosolenia calcar-avis (Schultze) Sundström 61

Pteropods 1, 168
Pterosperma sp. 86
Pyrodinium bahamense 90
Pyrodinium minimum 90
Pyrophacus horologicum Stein 1883 83
Pyrosoma atlanticum 175
Pyrosomida 175
Pyrrophyceae 72

Q

Quantitative analysis 25

R

Radial symmetry 47
Raphidophyceae 1, 84
Rhabdonella conica 117
Rhincalanus cornutus 159
Rhizosolenia 47, 48
Rhizosolenia crassispina Schroeder 60
Rhizosolenia cylindrus Cleve 60
Rhizosolenia imbricata Brightwell 59
Rhizosolenia robusta Norman 59
Rhizosolenia setigera Brightwell 59
Rhizosolenia spp. 21
Rhizostoma pulmo 129
Rhizostomeae 129
Rhopalophthalmus sp. 167
Rotifera 131

S

Sagitta enflata 136
Sagitta maxima 136
Salpa (Thalia) *dmocratica* 176
Salpingella attenuata 121
Sapphirina nigromaculata 163
Sarcodina 103
Schroederella delicatula (Peragallo) 53
Scyphozoa 178
Secondary production 37
Sedgwick rafter cell 28

Sedimentation chambers 24

Sedimentation volume 23

Semaeostomeae 128

Shannon - Wiener diversity index 43

Silicoflagellates 1

Similarity index 44

Siphonophora 124

Size determination 42

Size measurements 39

Skeletonema costatum (Greville) Cleve 51

Solmundella bitentaculata 124

Species diversity 43

Square net 30

Stage micrometer 39

Staining 21, 22

Stempel pipette 28

Stephanopyxis 47, 52

Stephanopyxis palmeriana (Greville) Grunow 52

Student net 11

Styrax 22

Subeucalanus flemingeri 147

Subsample (Aliquot) 25

Synchaeta sp. 131

T

Temora discaudata 151

Temora turbinata 152

Tentaculata 130

Thalassionema frauenfeldii (Grunow) Hallegraeff 67

Thalassionema nitzschioides (Grunow) Mereschkowsky 67

Thalassiosira 46

Thalassiosira eccentrica 49

Thalassiothrix longissima Cleve and Grunow 68

Thaliacea 175

Thaliaceans 175

Thallassiosiraceae 46

Thecosomata 168

Thenus orientalis 192

Tintinnids 105

Tintinnopsis acuminata 105

Tintinnopsis ampla 107

Tintinnopsis beroidea 107

Tintinnopsis cylindrica 108

Tintinnopsis directa 110

Tintinnopsis mortensenii 108

Tintinnopsis nordqvisti 109

Tintinnopsis radix 110

Tintinnopsis schotti 111

Tintinnopsis tocantinensis 108

Tintinnopsis tubulosa 110

Tomopteris helgolandica 135

Tortanus barbatus 158

Tortanus discaudata 157

Trachymedusae 124

Triceratium reticulum Ehrenberg 55

Trichodesmium erythraeum Ehrenberg 1830 87

Tunicates 205

Typical plankton net 10

U

Urochordata 205

Utermohl method 26

Utermohl tubular chambers 27

V

Velella velella 127

Volumetric methods 23

W

Water filtered 30

Wisconsin net 11, 12

Z

Zoea 192

Zooplankton 22, 24

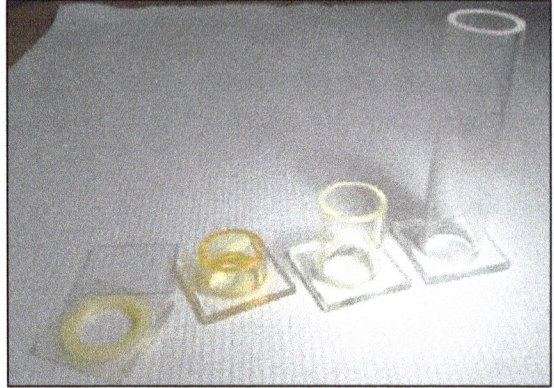

P. 24

Utermohl Tubular Chambers (P. 27)

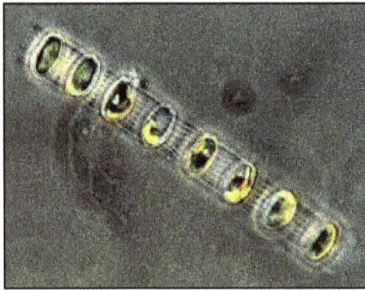

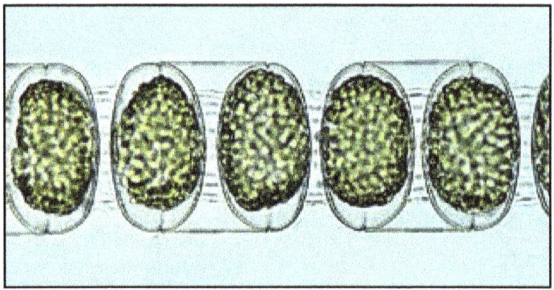

P. 52

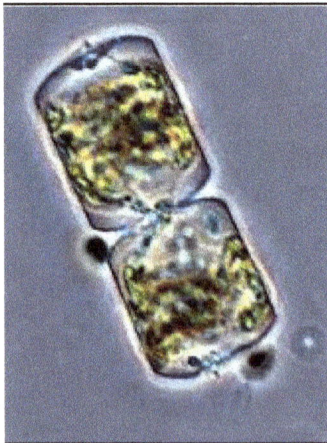

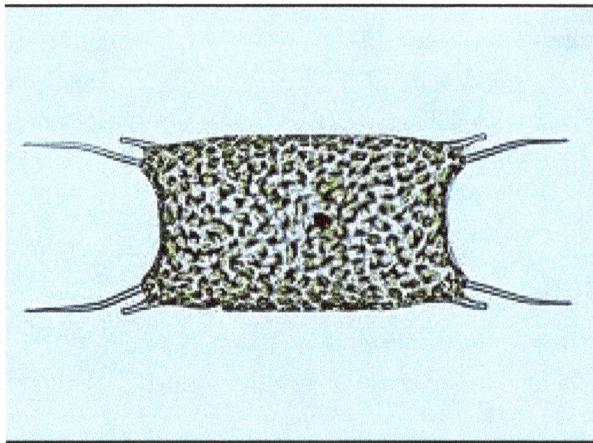

P. 54

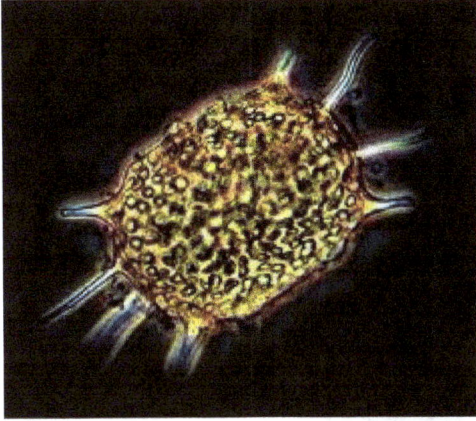
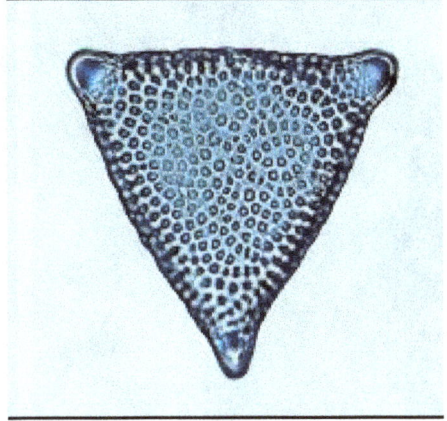

P. 55

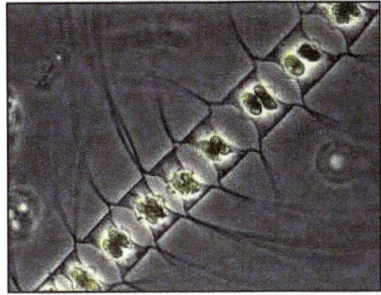
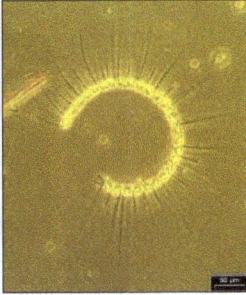
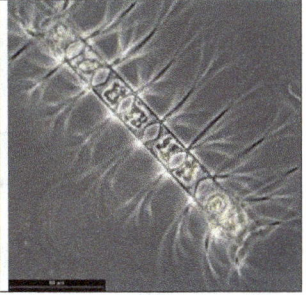

P. 64

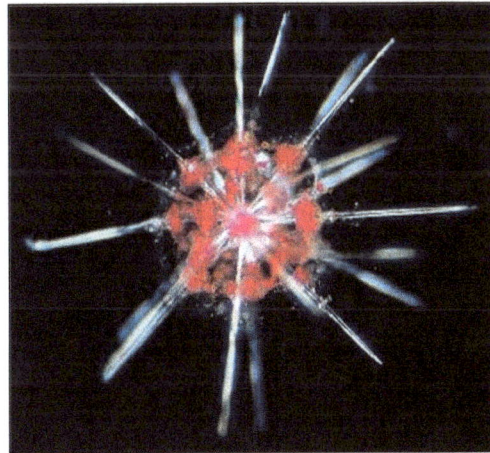

P. 104   **P. 123**

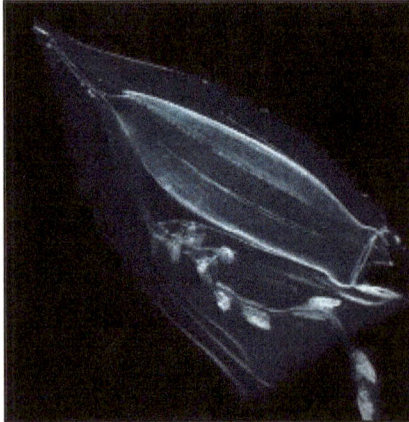

P. 125

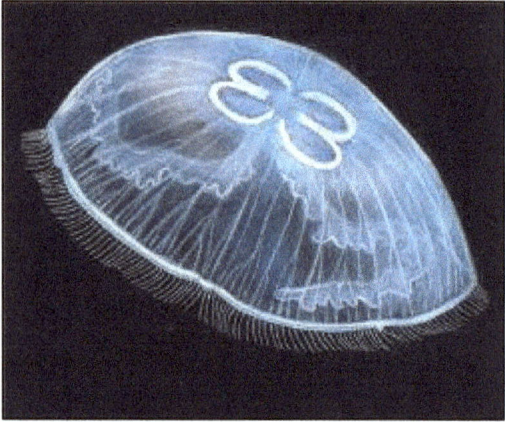

P. 128

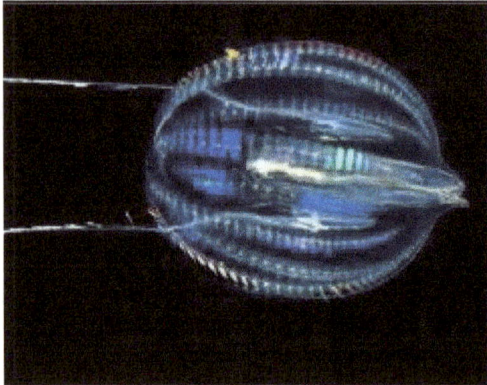

P. 130

P. 134

P. 135

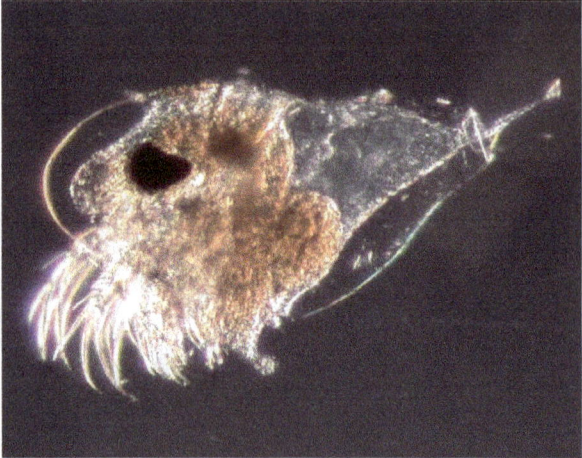

Evadne spinifera (P. 140)

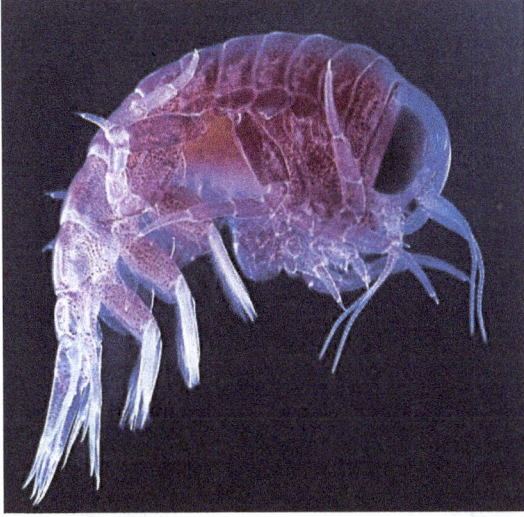

P. 142

P. 157

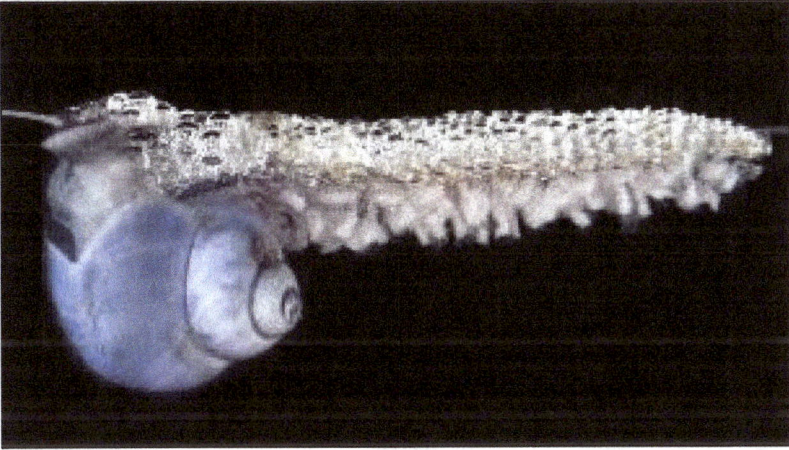

P. 168

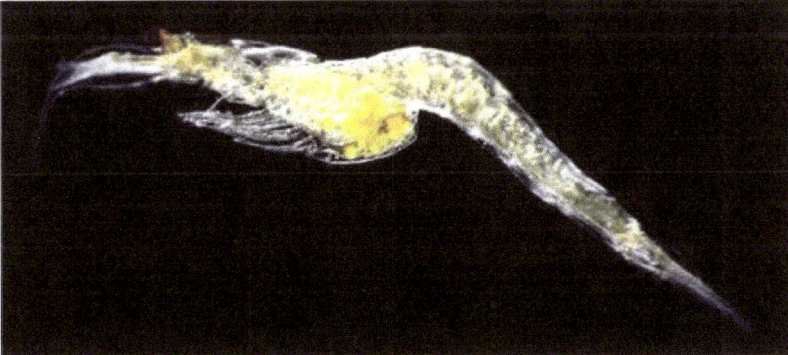

P. 167

P. 172

P. 172

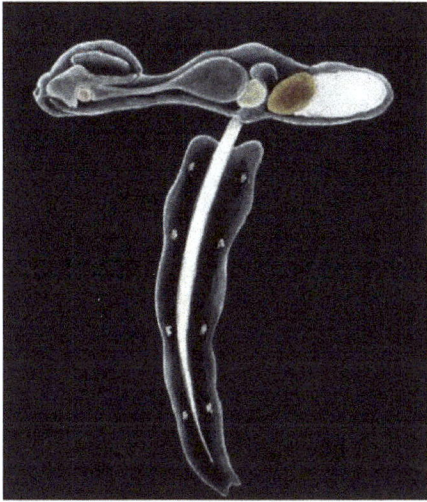
Frittilaria borealis (P. 175)

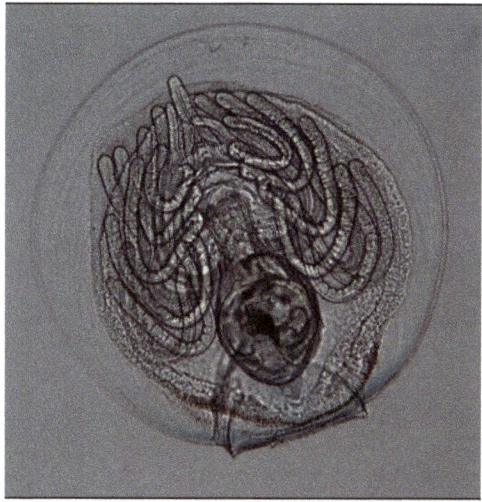
P. 181

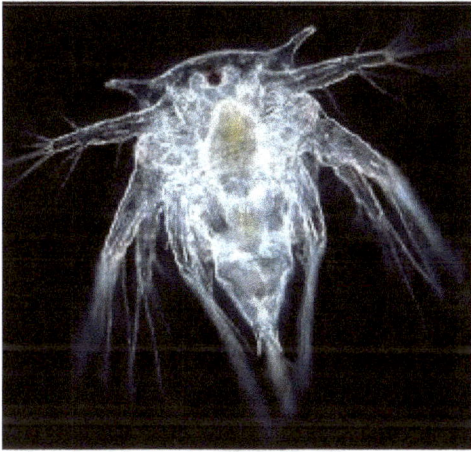
P. 190

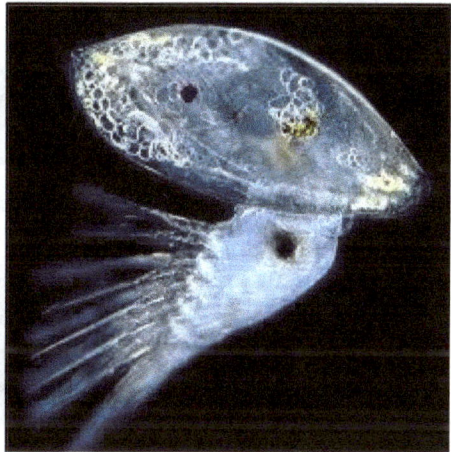
P. 191

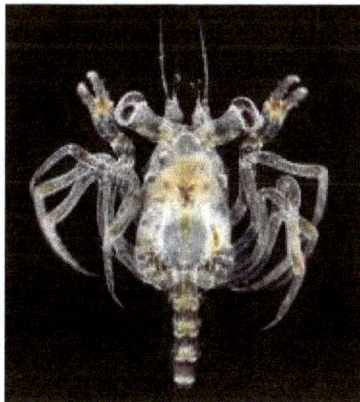
P. 193

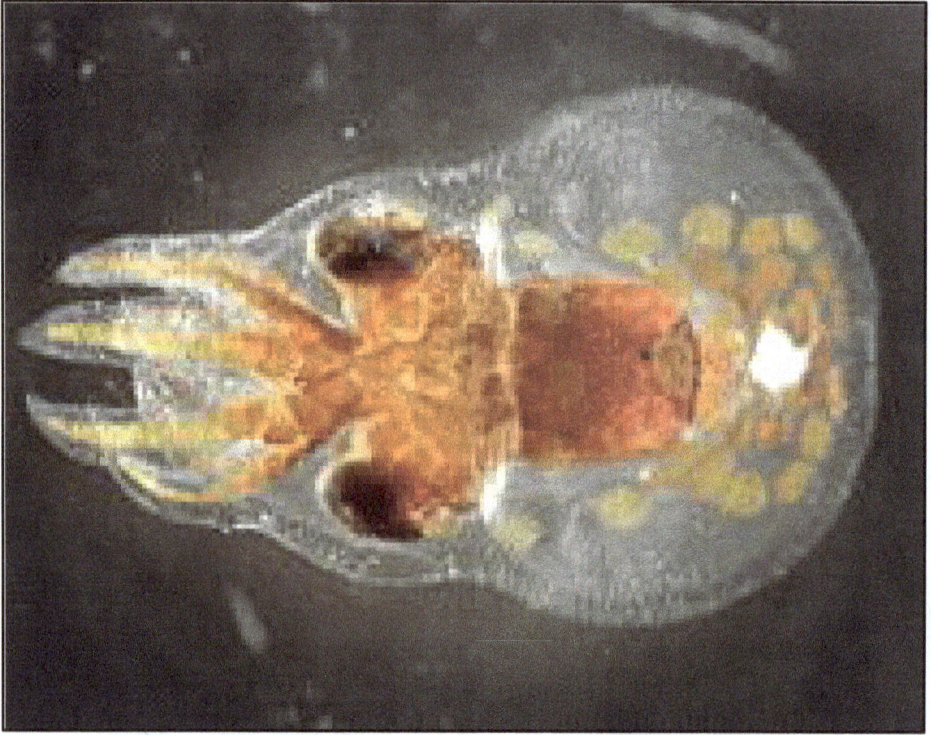

Paralarva of *Loligo* sp. (P. 197)

Advanced Paralarva of *Loligo* sp. (P. 197)

Hatchlings of *Octopus* sp. (P. 198)

Paralarva of *Octopus* sp. (P. 199)

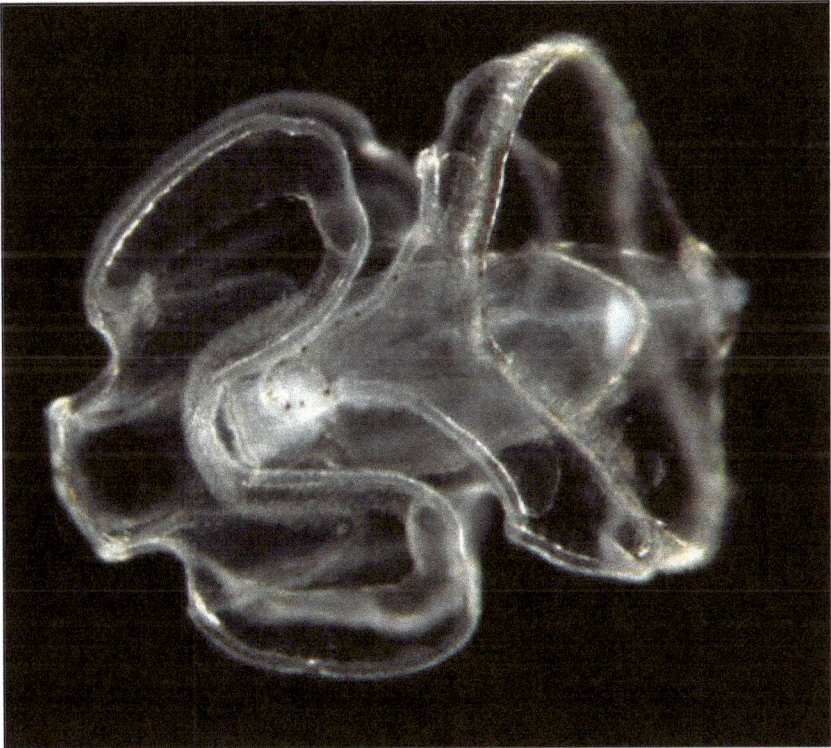

P. 205